Marine Electrical Basics Workbook

Fourth Edition

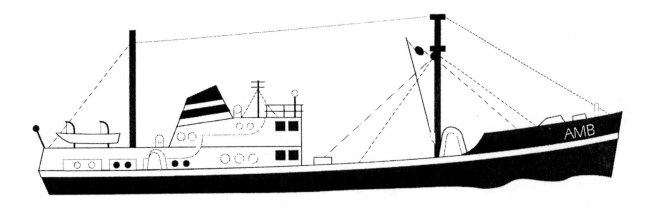

William A. Young

Government Institutes
Rockville, Maryland

T0308726

ABS Group Inc.

Government Institutes Division
4 Research Place, Rockville, Maryland 20850, USA
Phone: (301) 921-2300
Fax: (301) 921-0373
Email: giinfo@govinst.com
Internet: http://www.govinst.com

ISBN: 0-86587-681-9

Printed in the United States of America

This book is dedicated to my wife, Ann
(without whose gentle persuasion
this book would never have been written)

MARINE ELECTRICAL BASICS WORKBOOK

TABLE OF CONTENTS

LIST OF TABLES

LIST OF BLANK WORKSHEETS

LIST OF REFERENCES

1. *ABS Rules for Building and Classing Steel Vessels 1998-1999 Part 4 "Machinery Equipment and Systems* Copyright 1998 American Bureau of Shipping, ABS Plaza, 16855 Northchase Drive, Houston, Texas 77060 USA. *

2. *Recommended Practice for Marine Lighting Publication RP-12-97* Copyright 1997 Illuminating Engineering Society of North America (IESNA), 120 Wall Street New York, New York 10005

3. *IEEE Recommended Practice for Electric Installations on Shipboard* IEEE Std 45-1998 Copyright 1998 The Institute of Electrical and Electronics Engineers, Inc., 345 East 47th Street, New York, N.Y 10017 USA.

4. *NEC 1999 National Electrical Code* NFPA 70 Copyright 1998 National Fire Protection Association, Inc., One Batterymarch Park, Quincy, Massachusetts 02269 USA.

5. *46 CFR Parts 110 to 113 SUBCHAPTER J – ELECTRICAL ENGINEERING* Part of United States Coast Guard Shipping Regulations in the Code of Federal Regulations 46 CFR Parts 90 to 139 Revised as of October 1, 1997, U. S. Government Printing Office Washington D. C.

* Effective 1 January 2000, a complete nine-booklet reprint of the 1998-1999 ABS Rules for Building and Classing Steel Vessels was published. This new printing included a complete editorial revision and some technical revisions of part 4, which has been re-titled "Vessel Systems and Machinery." Part of this editorial revision was a re-numbering and re-organization of all section numbers. For vessels that must comply with the new rules (based on a contract date for construction after 1 January 2000), a matrix of all ABS rules used throughout this workbook is shown on the following page. In this matrix, the old rules are referenced to the new rules.

ABS RULES FOR BUILDING AND CLASSING STEEL VESSELS
1998-1999 EDITION VS. 2000 EDITION

Old 1998-99 Section	New 2000 Section	Book Module	Book Pages	Current Title
4/5.17	4-8-3/1.17 4-8-3/3.5 4-8-3/7.3.1	A4	A4-2	Ambient temperatures Rating Rating
4/5A1.1.2	4-8-1/5.1.1 4-8-1/5.1.6	C11, C14	C11-1, C14-1	One line diagram Other information
4/5A2.1.2	4-8-2/3.1.1	B8	B8-2	General
4/5A3.3	4-8-2/5.5	C11	C11-7	Emergency services
4/5A4.1.4	4-8-2/7.7.4	B10	B10-4	Motor control center feeder
4/5A4.1.5	4-5-1/5.5.2 4-8-2/7.7.1(e) 4-8-2/7.7.6	A4	A4-1	Electrical circuits Minimum conductor sizes Motor branch circuit
4/5A4.1.6a	4-8-2/3.7	B8	B8-2	Transformers and converters - Continuity of supply
4/5A4.1.6b	4-8-2/3.7	B8	B8-2	Transformers and converters - Arrangements
4/5A4.1.7	4-8-2/7.15	C11	C11-8	Ventilation system circuits
4/5A4.1.9	4-5-1/5.5.2	C11	C11-9	Electrical circuits
4/5A5.1.3c	4-8-2/9.13	A4, A5, B7	A4-1, A5-2, B7-2	Protection of feeder cables
4/5A5.13.1	4-8-2/9.17.1	B10	B10-4	Motor branch circuit protection
4/5A5.13.4	4-8-2/9.17.2 4-8-3/5.7.1	A4	A4-1	Motor overload protection Overload and under-voltage protection
4/5A5.13.5	4-8-2/9.17.3 4-8-2/9.17.4 4-8-3/5.7.1	B10	B10-1, B10-2	Under-voltage protection Under-voltage release protection Overload and under-voltage protection
4/5A5.15.1	4-8-2/9.19.1 4-8-2/9.19.2	B8	B8-1	Protection at primary side only Protection at both primary and secondary sides
4/5A5.15.2	4-8-2/9.19.2	B8	B8-2	Parallel operation
4/5A6.1	4-3-4/5.1 4-3-4/11.1 4-8-2/7.11	C11	C11-8	System arrangements Power supply feeders Steering gear power supply feeders
4/5A6.3	4-8-2/9.17.5	B10	B10-2	Protection of steering gear circuits
4/5A6.3.2	4-3-4/11.5	B10	B10-2	Under-voltage release
4/5A7.1.4	4-8-2/9.21	B6	B6-2	Protection of lighting circuits
4/5A7.3.3	4-8-2/11.3.2	C11	C11-9	Branch circuit

Old 1998-99 Section	New 2000 Section	Book Module	Book Pages	Current Title
4/5A10.1.1	4-8-2/7.15 4-8-2/11.9.1 4-8-4/13.2	C11	C11-9	Ventilation system circuits Ventilation systems Power ventilation
4/5B1.1	4-8-1/5.3.1	C15	C15-1	Booklet of standard wiring practice
4/5B2 Table	4-8-4/21.11	B8, C15	B8-6, C15-3	Cable bending radius
4/5B2.13.2	4-8-4/9.3	B10	B10-4	Disconnecting arrangements
4/5B3.1.2	4-8-2/7.7.1(a) 4-8-4/21.1.3	A4	A4-2	General Choice of insulation
4/5B3.1.3	4-8-2/7.7.1(d) 4-8-4/21.5 4-8-4/29.11	C14	C14-1	Voltage drop Cable voltage drop Voltage drop measurement
4/5B3.1.4	4-8-4/21.1.2 4-8-4/21.9.1	C15	C15-1, C15-2	Restricted locations General
4/5B3.5.3	4-8-4/21.1.5	C15	C15-2	Signal cables
4/5B3.9.1	4-8-4/21.9.2 4-8-4/21.9.3 4-8-4/21.9.3(a) 4-8-4/21.9.3(b) 4-8-4/21.9.4	C15	C15-4	Spacing for cable support Clips, saddles, straps Sizes Non-metallic material Non-metallic conduits
4/5B3.11.2	Requirement removed	C15	C15-2, C15-4	Clearance and Segregation
4/5B3.13	4-8-4/21.13 4-8-4/21.13.1 4-8-4/21.13.3 4-8-4/21.13.4	C15	C15-5	Deck and bulkhead penetrations General Non-watertight penetrations Collision bulkhead
4/5B3.15.1	4-8-4/21.15.1	A4	A4-2	General
4/5B3.17.1	4-8-4/21.17	C13	C13-2	Emergency feeders
4/5B3.33	4-8-4/21.25	C11	C11-7	Installation of cable junction boxes
4/5B7.1.2	4-8-4/27.11	B6	B6-2	Lighting circuits in hazardous areas
4/5C3.3.3	4-8-4/5.1.3	C13	C13-1	Accumulator batteries
4/5C4.17.2	4-8-3/5.7.2 4-8-4/9.3.1	A4, B10	A4-1, B10-1	Disconnecting means General
4/5C4b Table	4-8-4/5.1.3	C13	C13-1	Accumulator batteries
4/9.61.2	4-4-1/11.3 4-4-1/13.3.4 4-4-1/15.2 4-6-4/13.9.1 4-6-5/3.3.1 4-8-2/11.9.3	C11	C11-9	Manual emergency shutdown Remote shutdown Emergency shutdown Fuel oil transfer pumps Service and booster pumps Forced-draft fans

Revised 6 January 2000

PREFACE

During my years in the field of Naval Architecture and Marine Engineering, I have discovered there is precious little in the way of formal instructions for electrical marine design and installations.

Because the field is so specialized, and experienced marine electrical designers and installers are hard to come by, it is frequently necessary to hire inexperienced people and train them.

For this purpose this workbook is written. This workbook has been devised as a series of 15 Lessons divided into 3 modular segments to take the beginner step-by-step from basic ship construction through building and installing a cable wireway.

The Lessons include many diagrams, charts, formulas, examples, and solutions in addition to giving simple problem solving exercises. The lessons had to be in a simple enough format to allow the novice to design and install approved power and lighting systems with all the various calculations involved.

Since these lessons had been proven in actual use for many years, they were first compiled into a workbook in 1977 and used to teach novices. They were originally set up with an instructor in mind, but can be followed by the inexperienced individual for self-teaching.

Even high school graduates with little or no formal training can take part of the burden of design and installation from the engineer's shoulders.

This workbook does not preclude the designer or installer from the responsibility of being familiar with the requirements of the American Bureau of Shipping, U.S. Coast Guard, and IEEE-45 regulations, but rather aids in their understanding by putting these rules in a more comprehensible form.

The first three lessons in Module A of this workbook are intended to familiarize the novice with basic ship construction and terminology before proceeding with the design of power and lighting systems in modules B and C.

Various study guides have been included as appendices to aid and enhance the lessons that require them.

Blank worksheets have been provided in various lessons throughout this workbook for the designer and installers use. These work sheets may be freely copied, or used as a guide for making your own. I have found that having proper data listed speeds up design and greatly improves the accuracy of the final installation.

William A. Young

ABOUT THE AUTHOR

William A. "Bill" Young is a Program Supervisor for Quality, Safety, and Environmental systems in the Marine Division of the American Bureau of Shipping (ABS) Group, Inc. with headquarters in Houston, Texas. A member of the American Society for Quality, Bill has over 35 years experience in the Marine Industry with the majority being in electrical design and installations.

After graduating from high school, he started his marine career in the U.S. Navy in June of 1958, entering into their communications and electronics field. Bill has worked extensively with marine, industrial, and commercial electrical designs and installations, and holds a current Master Electricians License.

Bill resides in Santa Fe, Texas with his wife of 35 years. They have five daughters and nine grandchildren.

INTRODUCTION

A ship is a floating city. It includes within its hull not only propulsion and power generation, but all the equipment and systems needed to sustain a crew and passengers, and deliver a cargo or service. A vessel must provide food, accommodations, and sanitation service for all on board. Unlike shoreside installations, there is no availability of unlimited power sources for future growth such as in municipal power and lighting companies, or community water and sewage service. Once a vessel leaves the shelter of a port, it is no longer dependent on shore facilities and must be designed to be totally self-sufficient.

Fuel is costly and must be carried on board the ship. Over designing on the part of the ships systems results in over sizing of the ships generators, resulting in increased fuel costs. Weight on ships is a critical factor, therefore power and lighting system calculations must be concise to prevent using cables and equipment which are unnecessarily oversized.

Ships electrical systems must be designed to accomplish the most efficient distribution and use of power without excess allowances for future growth. Calculations of electrical lighting and power systems must be concise, not only to carry the amount of power required, but to keep the weight of the electrical system as low as possible.

I.   Before any electrical designing or installing can be done for a vessel, it is imperative to have a general knowledge of how a vessel is constructed, and how this is portrayed on structural drawings.

   A. It is nearly impossible to decipher the background for lighting and power system plans unless you can understand the marine terminology usually shown on General Arrangement drawings.

   B. A wireway to carry electrical cables cannot be correctly detailed unless the abbreviations on structural plans can be interpreted.

   C. Interferences cannot be avoided unless a general understanding of abbreviations and symbols used in the marine field is known.

   D. Even Noah did not build the Ark, which was the earliest recording of ship construction, without exact directions.

II. Although this workbook is not intended to instruct on the aspects of building a ship, the purpose of the first three lessons is to help you recognize the many words which have special abbreviations for use on Marine Drawings.

   A. Most of this information can be found in basic shipbuilding books.

   B. It is not the intention to have you memorize all the terms included, but once these basic lessons are filled out, they will give you a ready reference for future use.

MARINE ELECTRICAL BASICS

WORKBOOK

Module A

FUNDAMENTALS

I.   The usual mechanical drawing conventions apply in the marine field, with minor exceptions. There are, in addition, a number of special rules peculiar to the Naval and Marine fields.

    A. Vertical lettering is universally used in hull and electrical drawings and usually used in machinery drawings. UPPER CASE LETTERS are standard.

    B. In profiles, backgrounds, arrangements or details, the bow is always to the right.

    C. For hull drawings normally it is necessary to draw only one side of the vessel (in plan or sections) since ships are largely symmetrical. Electrical backgrounds require both sides.

        1. In deck plans it is usual to show the port side simply because it is easier to draw a curve in that direction.

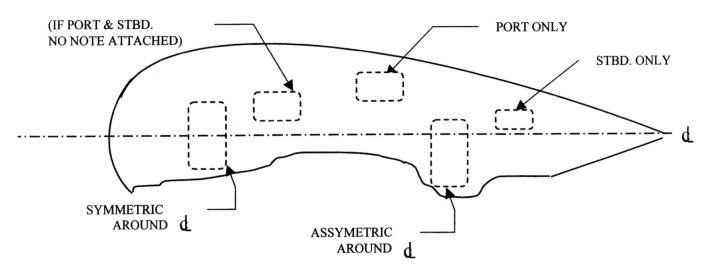

"Port side shown-starboard side to opposite hand, except as shown or noted"

FIGURE A1-1

    D. Decks and arrangements are normally drawn looking down (called a plan view). A view looking up, when used, is called a reflected view.

    E. Longitudinal members are drawn with the forward end to the viewer's right hand and are considered as being viewed from starboard. If member is duplicated P/S (port and starboard), only one side need be shown.

    F. Transverse bulkheads, etc., are normally symmetric and again only one side need be drawn. The choice of side varies in different drawing offices but is usually either:

1. Starboard side, viewed <u>from</u> ⊗ (amidships) or;

2. Port side, looking <u>towards</u> ⊗ (amidships).

    (a) A little thought will show that either of the above makes easy use of the body plan.

II.   The usual coordinates for locating things on the ship are:

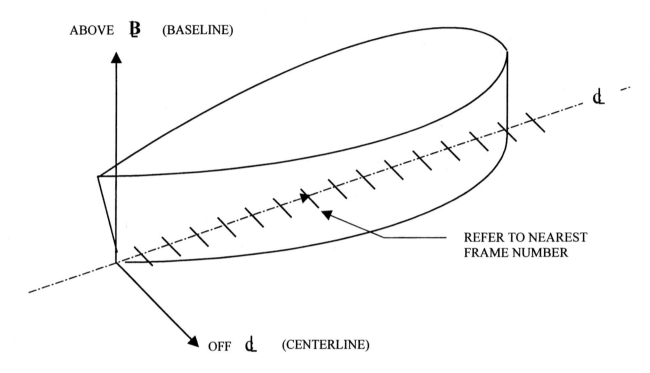

FIGURE A1-2

A. Frames are <u>located</u> starting with the transom frame at the stern. This is the only spot in the ship where the location of internal framing bears any direction relationship to the external hull form (we do <u>not</u> need a frame exactly at the ⊗ (midships) or the forward perpendicular or any of the other imaginary stations). Having located the transom frame, the others are located mechanically by use of appropriate spacing.

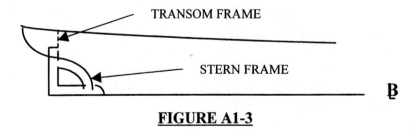

FIGURE A1-3

B. In the United States frames are usually <u>numbered</u> starting at the bow. European practice, which is just the opposite, makes more sense since the first frame located is at the stern and we read from left to right anyhow.

C. Usual scales are:

For decks, etc.:	1/4″ = 1 ft.
For transverse bulkheads:	1/2″ = 1 ft.
(Simply to fill width of 30" drafting paper)	
For Isometric Backgrounds:	1/8″ = 1ft.

D. At the scale of ¼" = 1 ft., a plate ½" thick would, if accurately drawn in section, be ½ ÷ 48" = 1/96" wide. Since it is impractical to try to cross hatch such a narrow band, we simply draw sections of plates and shapes as single heavy black lines and do not pay too much attention to exact thickness.

III. A convenient location method is the use of imaginary 12" panels on a drawing:

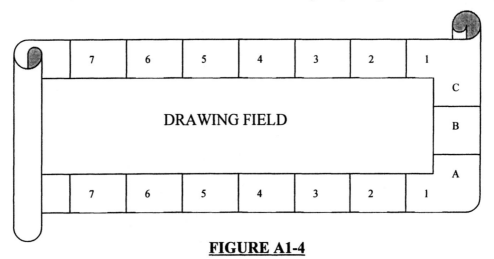

FIGURE A1-4

A. The innumerable details and/or sectional views are labeled with a number locating their panel, and save the trouble of searching the entire drawing.

B. Horizontal panels using letters of the alphabet are also sometimes used to give the latitude as well as the longitude for locating details.

IV. Details should not be shown on more than one drawing. For example, a bracket connects a deck beam and a side shell frame. The beams and frames are shown on two different drawings. Show the bracket in both details, but identify and detail it only in one. Otherwise, when alterations come through one will be changed but not the other, or the mold loft will lay off the same bracket twice, etc.

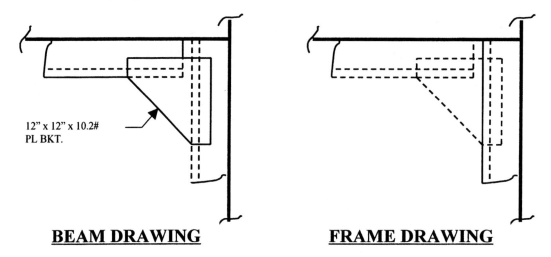

12" x 12" x 10.2#
PL BKT.

BEAM DRAWING   **FRAME DRAWING**

FIGURE A1-5

A. Each drawing should contain a list of reference drawings. These would include the preliminary design drawing on which based, plus production drawings of all adjacent structures.

B. Owing to the great length of ships, it is impractical to try to show a shell or deck drawing for its entire length.

(A 600 foot ship at ¼" = 1 ft. would require a deck drawing about 13 or 14 feet long)

To get around this, it is usual to split the ship into two or three arbitrary sections for drawing purposes. These always stay the same for a given design.

C. Two convenient methods of showing a sectional view are:

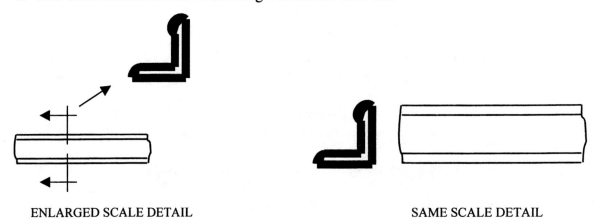

ENLARGED SCALE DETAIL   SAME SCALE DETAIL

FIGURE A1-6

D. In the case of angles of unequal legs, the arrow head lands on the leg specified first. For example:

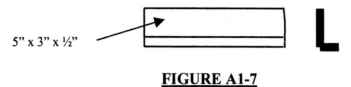

FIGURE A1-7

E. Welded joints are located by the symbol: **$**

Examples:

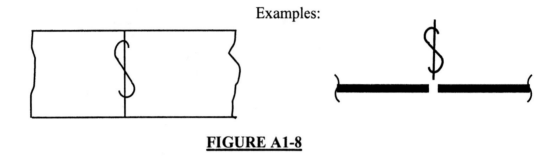

FIGURE A1-8

F. Where a group of members have something in common, such as scantling or spacing, they may be grouped for notation:

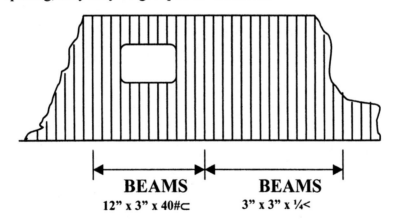

BEAMS   **BEAMS**
12" x 3" x 40#⊏   3" x 3" x ¼<

FIGURE A1-9

V. Most drawings are not intended for scaling-off. They are not drawn with super-accuracy in the first place and frequently change size in reproduction.

A. The molded surface for various structures is as follows:

MEMBER	MEASURE FROM
Shell plating:	Inside surface
Longitudinal members on centerline:	Down the center of the member
Longitudinal members not on centerline:	Inboard surface
Tank top, decks, etc.:	Underside
Transverse bulkheads, floors:	Midship side
Deck house sides:	Outboard surface

B. An underlined dimension (|◄—— <u>18"</u> ——►|) indicates that part of the drawing is not drawn correctly to scale.

C. Dimensions less than two feet are given in inches and fractions.

D. Lettering of notes, dimensions, scantlings, etc., are generally about 1/8" high.

MARINE EXAMINATION

The following exam is intended to ensure you have a basic understanding of marine drafting terms, and to provide you with a future source of reference. Exams will not be graded but will be discussed to allow correcting any answers.

1. Where does the first record appear of the art of shipbuilding?

    _____

2. Define the following:

    (a) ₫   _____

    (b) ₽   _____

    (c) ABL   _____

    (d) AP   _____

    (e) FR.   _____

    (f) FP   _____

    (g) CVK   _____

    (h) ATHW.   _____

    (i) BPMH   _____

    (j) RWTH   _____

    (k) LBP   _____

    (l) LOA   _____

3. TRUE or FALSE?

    (a) _____ Athwartships means run in a fore and aft direction.

    (b) _____ Transverse means run in a port and stbd. direction.

    (c) _____ A stiffener is an angle, flat bar, or beam, welded to a deck or bhd. for strength.

    (d) _____ The stern of a ship is the furthest point aft on a ship.

(e)   _____   The forward perpendicular is the furthest point fwd. on a ship.

(f)   _____   A longitudinal bulkhead would run from the port side of a ship to the stbd.

(g)   _____   The freeboard deck is normally the uppermost complete deck of a ship that is exposed to weather and sea.

(h)   _____   The auxiliary machinery on a ship is the most important for the vessel to run.

(i)   _____   A ballast tank on a ship is a compartment for storing fluorescent fixture parts.

(j)   _____   Binnacles are deposits on the hull of a vessel under the waterline caused from too much time in the water without being dry-docked and the hull scraped.

(k)   _____   Chain lockers were used in old times for storing slaves while crossing the ocean.

4. Which definition applies to each of the words below?

   (a)   Swash Bhds.   _____

   (b)   Stanchions   _____

   (c)   Skeg   _____

   (d)   Appendage   _____

   (e)   Shell Plates   _____

   (f)   Base Line   _____

   (g)   Camber   _____

   Definitions

   A. The after part of the keel, extending below the shell plating, to give support for the stern post to rest upon.
   B. The transverse curvature of the deck.
   C. Outermost steel plating of a ship.
   D. Short columns or supports for decks, hand rails, etc.
   E. Straight horizontal line at the very lowest line of the molded surface on a ship.
   F. Small portions of the ship which extend beyond the shell plate.

   G. Bhds. built to prevent liquid from gaining momentum in flow from one end of a
      tank to another.

5. In shipboard terminology, describe in one word the following:

   (a)   Wall   _____

   (b)   Window   _____

   (c)   Width   _____

   (d)   Chimney   _____

   (e)   Liquid storage room   _____

   (f)   Room or spaces   _____

   (g)   Ceiling   _____

   (h)   Floor   _____

   (i)   Stairway   _____

6. From the definitions below, what do the following lines represent on a drawing?

   (a)   ━━━━━━━━━   _____

   (b)   ─────────   _____

   (c)   ▬ ▬ ▬ ▬ ▬   _____

   (d)   ──╲╱──╲╱──   _____

   (e)   ───╲_────   _____

   (f)   ─ · ─ · ─ · ─   _____

   (g) |◄─────►| _____

   (h)   ─ · · ─ · · ─   _____

   <u>Definitions</u>

      A. Expanded metal bulkheads, or hidden plates.
      B. Short breaks in a section.
      C. Dimension lines.

       D. Phantom line.

       E. Cables.

       F. Long breaks in a section.

       G. Centerline.

       H. Background.

7. What does this symbol stand for:

     ⌀   _____

8. Draw these structural shapes:

    (a)   CHANNEL _____

    (b)   "T" BEAM _____

    (c)   ANGLE   _____

    (d)   "I" BEAM   _____

9. When cutting a hole in a steel plate, beam, or girder, compensating steel means to: ____

       A. Cover the cutout with a larger piece of steel.

       B. Add a stiffener on each side of the hole.

       C. Replace the amount of steel removed.

       D. Make the hole smaller than necessary.

10. What is the normal voltage for a sound powered telephone?

11. What colors are the port and stbd. navigation lights?

    (a)   Port _____

    (b)   Stbd._____

For the sake of space many marine drawings resort to standard abbreviations instead of spelling out the entire word. In order to more fully understand the drawing you are looking at, whether it be structural, piping, mechanical, or electrical, you should have a firm understanding of marine abbreviations. This lesson is intended to provide you with a ready reference to marine abbreviations.

Lesson 2

From information learned in Lesson 1, and definitions listed in Appendix A, fill in as many of the following common abbreviations as you can. Any remaining will be filled in during open discussion.

-A-

ABL –

ABS –

ACCOM. –

A & D –

AE –

AF –

ANT. –

AP –

ARRGT. –

AT –

ATHW. –

AUX. -

-B-

BALL. –

BHD. –

BKT. –

Lesson 2 – Abbreviations and Initials on Marine Drawings

BL –

B/M –

BPMH -

-C-

CB –

CC –

CD –

CL –

COMP. –

COMPR. –

CONS. –

CONT. –

CONT'D. –

CONTR. –

CSG. –

-D-

DETS. –

DISCH. –

DK. –

DO –

DRNS. -

-E-

E & IWD –

ELEM. –

ELEV. –

EMB –

EMERG. –

ENG. –

EP –

ER –

EXH. –

EXT. -

-F-

FDN. –

FDRS. –

FLA –

FO –

FP –

FR. –

FW –

FWD. –

-G-

GEN. -

-H-

HP -

-I-

IB –

IEEE –

INBD. –

INST. –

INT. –

ISOM. -

-J-

JB -

-K-

KP -

-L-

LBP –

LCG –

L/M –

LO –

LOA –

LONG'L. –

LT –

Lesson 2 – Abbreviations and Initials on Marine Drawings

LTG. –

LVP –

LVR -

-M-

MACH'Y. –

MAG. –

MFLD. –

MISC. –

MNS. –

MODS. -

-N-

NAV. –

-O-

OT –

OUTBD. –

OVBD. –

OVHD. -

-P-

PL –

P/L –

PLBG. –

-R-

RED.V. –

RPM –

RWTH -

-S-

SHP –

SPT –

SR –

ST –

STAN. –

SW -

-T-

TK –

TRANS. –

TRK. –

T/S –

TT -

-U-

USCG -

-V-

VCG –

VENT. –

VERT. –

VL -

-W-

WD –

WL –

WT –

WW -

-X-

XFER. –

XFMR. -

MARINE CROSSWORD

TEST YOUR KNOWLEDGE OF BASIC SHIPBUILDING TERMS by completing this crossword puzzle. Where an abbreviation (abb.) is called for; the abbreviation of the answer, not the question, should be filled in.

ACROSS

1. The curvature of a deck toward the center for shedding water.
3. List made up of quantity and kind of material needed (3 words).
9. The line perpendicular to the base line and intersecting the bow at the design water line (abb.).
10. Sheet metal conduit connecting the boiler smoke box with the base of the smokestack.
13. Board extending across a rowboat for stiffening and seat.
15. Ships hallway.
16. American _ _ _ _ _ _ of Shipping.
17. A shape usually having two projecting horns for belaying a rope.
18. Left side of a vessel.
21. An access from a weather deck protected by a hood from sea and weather (2 words).
24. Vertical plane running lengthwise through the ship at the bottom to divide it into port and stbd. halves (abb.).
26. Piece of material let into a slack place to fill out a fair surface.
30. The line perpendicular to the base line and intersecting the aft stern post (abb.).
31. Length between _ _ _ _ _ _ _ _ _ _ _ _ _.
33. Type of joint in a pipe permitting linear movement to take up expansion and contraction due to temperature changes (abb.).
34. Pipe to open air for use when filling or emptying a tank (abb.).
36. Term used to designate the transverse ribs that make up the skeleton of a ship (abb.).
37. Table of deck heights and sight edges and other information to fix the entire molded form of a ship.
40. Part of the contents of sea water.
41. The aft end of a vessel.
42. Outside plating of a ship.
44. Long, cylindrical, heavy forging connecting engine and propeller.
45. Plate adjoining the hawse pipe to prevent the chafing of the anchor against the ships bow.
47. A sign, usually plastic, or metal, to designate the function of a piece of equipment (abb.).
48. Ships couch.
49. Source of electrical power.
52. Locker for storing anchor chains.
55. Bottom line of a ship from which all vertical dimensions are given (abb.).
56. Rescue boat (abb.).
57. _ _ _ _ _ ship weight.
58. Color of a starboard running light.
59. Class of bulkhead impervious to water (abb.).

ACROSS (Continued)

61. Cable trough (abb.).
62. Crew bedroom.
64. Vertical shaft formed by bulkheads.
68. Clean linen _ _ _ _ _ _.
70. _ _ _ _ axial flow vent fan.
71. That part of a ship at, or adjacent to, the bow (abb.).
72. Inbd. To _ _ _ _ _ (abb.).
73. Stairway.

DOWN

2. Amount of rotations in 60 seconds (abb.).
3. Magnetic compass, usually on the wheelhouse top.
4. Man who lays out the ships lines, full size, for templates.
5. The palm of an anchor.
6. Radio aerial (abb.).
7. Across the ship, usually for passages.
8. Pole serving as a mast, boom gaff, bowsprit, etc.
11. Means of emergency egress (abb.).
12. Across the ship, usually for bulkheads and beams.
14. Same as 64 across.
16. Heavy steel castings for securing lines, usually fitted to the deck.
19. Over the side (abb.)
20. Drains from decks to carry off water.
21. Type of plug screwed into the bottom of a ship to provide drainage from compartments when in drydock..
22. Term used to describe power from a motor (abb.).
23. Small room adjacent to bridge for determining vessels course.
25. Channel burned out by cutting torch.
27. Angle used to strengthen bulkheads or decks.
28. Center of gravity from bow to stern (abb.).
29. Mark painted on ships side designating depth to which vessel may be loaded.
32. Handle on a watertight door used to hold door shut.
34. Opposite of forward (abb.).
35. The width of a ship.
37. _ _ _ cooled bearings.
38. Room with toilet and shower (abb.).
39. Pipe used for determining amount of liquid in a tank (abb.).
42. Bow frame forming the apex of the triangular intersection of the forward sides of a ship.
43. Dimension between forward and aft perpendiculars (abb.).
44. Bulkheads in tanks fitted to decrease the sloshing action of liquids.
46. Color of the port running light.
47. Floor covering in solid sheet form, usually imprinted with patterns (abb.).

<u>DOWN</u> (Continued)

49. Ships kitchen.
50. Burned engine gas (abb.).
51. Water _ _ _ _ bulkhead.
53. Changes to drawings (abb.).
54. Space between the inner and outer bottom skins of a ship, usually from fwd. to aft. (abb.).
55. Forward part of a ship.
60. _ _ _ and running lights (abb.).
63. Casting or chest containing several valves (abb.).
65. Main center pillar posts of a ship (abb.).
66. To the inside (abb.).
67. To change from one spot to another (abb.).
69. Outside (abb.).

MARINE CROSSWORD

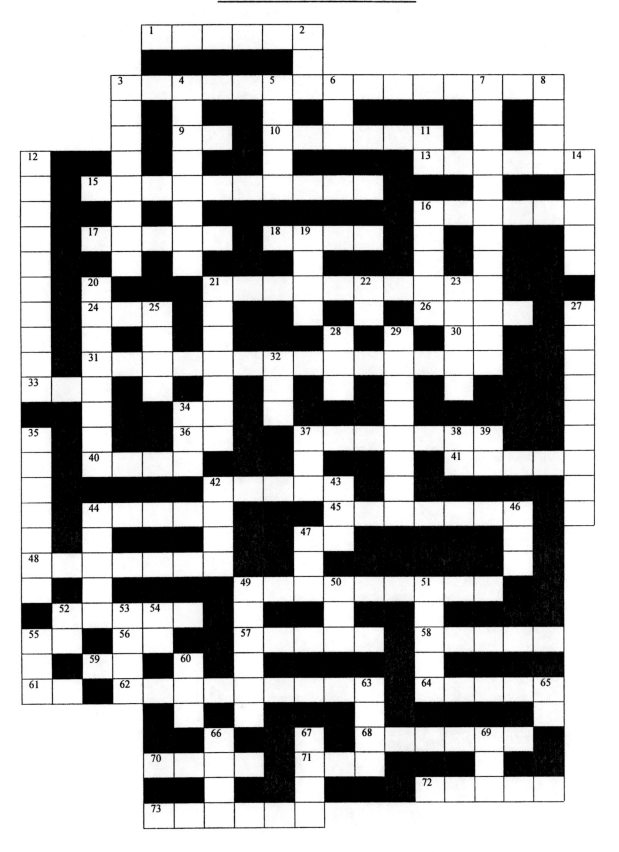

Before any cables can be selected, or shown on a drawing for shipyard installations, and before any power panels can be ordered, it is necessary to calculate the cable and circuit breaker sizes based on the characteristics of each individual motor.

I.　　Diagram of a typical motor circuit as shown on a One-Line Electrical Diagram:

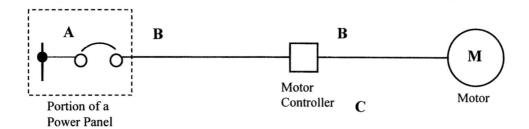

Lesson 4

FIGURE A4-1

II.　　The U. S. Coast Guard in Subpart 111.70-1(a) [Reference 5] requires that all motor circuits meet ABS rules [Section 4/5 in Reference 1]. ABS Rules for sizing circuit breakers and cables for motors over 0.5 KW or 6 Amps are:

A.　The Circuit Breaker (A) must be rated at least 115% of F.L.A. (Full Load Amps) [ABS Rule 4/5C4.17.2]. If the 115% does not correspond to a standard size, the next higher rating may be used, but not over 150% of the allowable current carrying capacity of the conductor. [ABS Rule 4/5A5.1.3c].

B.　The cable (B) must be rated to carry at least 100% of the motor F.L.A. [ABS Rule 4/5A4.1.5]. In addition, a motor branch circuit cable cannot be smaller than 1.5mm^2 (or 14 AWG) [ABS Rule 4/5A4.1.5]. The cable chosen cannot have a current carrying capacity which is below the rating of the circuit breaker chosen [ABS Rule 4/5A5.1.3c].

C.　The motor controller (C) must have motor protection (overloads) between 100% and 125% of the motor F.L.A. [ABS Rule 4/5A5.13.4].

III.　　**EXAMPLE NO. 1:**

Question 1: If you had a 440 Volt – 3 Phase motor rated at 200 Horsepower, what would be the circuit breaker Frame and Trip Size required?

Question 2: What would be the cable size required using IEEE-45 approved cables?

SOLUTION NO. 1

A.　First find the watts: (1 horsepower at 440Volts-3 phase is equal to 746 watts).

Therefore, a 200 horsepower motor equals: 200 x 746 = 149,200 watts.

B. Next, using the formula for 3 phase motors shown below, find the motor Full Load Amps (F.L.A.):

$$I = P \div E \times 0.8 \times \sqrt{3}$$

Where: I = Current in Amps
P = Power in Watts
E = Electromotive Force in Volts
0.8 = Assumed motor power factor for vessels
$\sqrt{3}$ = 1.732

$I = 149200 \div 440 \times 0.8 \times 1.732$

$I = 149200 \div 609.664$

$I = 244.73$ F.L.A. (Round off to 245 Amps).

C. To size the circuit breaker (A) multiply the motor F.L.A. times 115%:

$245 \times 1.15 = 281.75$ Amps (Round off to 282 Amps).

D. Using the enclosed Table A4-3, we see that the closest circuit breaker <u>Trip</u> size above 282 Amps is 300 Amps, therefore:

Answer 1: The circuit breaker must be 400 Amp Frame / 300 Amp Trip.

E. The cable must carry at least 100% of the motor F.L.A. of 282, but also must not be sized below the rating of the circuit breaker trip, which is 300 Amps.

F. Ambient temperatures in engine and boiler rooms are assumed to be 45^0 C (113^0 F) [ABS Rule 4/5.17]. The rated operating temperature of the cable insulation is to be at least 10^0 C (18^0 F) higher than the ambient temperature expected to exist [ABS Rule 4/5B3.1.2], resulting in a minimum cable insulation rating of 55^0C (131^0F).

G. IEEE-45 table 9-1 lists cables with insulation ratings starting at 75^0C. To simplify the learning process, and for the purpose of this class, it will be assumed that all motors are in engine rooms, and all cables used will have insulation rated at 75^0 C (167^0 F). In addition, cables subject to mechanical damage must be provided with braided metallic armor [ABS Rule 4/5B3.15.1]. For the purpose of this class, all cables will be selected with bronze armor for protection.

H. Using the enclosed Table A4-4, pick the three conductor bronze armored cable that is the nearest to, but not below, 300 Amps with an insulation rating of 75^0C.

Answer 2: The cable must be a TTNB-500 which is rated at 329 Amps.

IV.   <u>EXAMPLE NO. 2</u>:

Question 1: If you had a 440 Volt – 3 Phase motor rated at 15 Horsepower, what would be the circuit breaker Frame and Trip Size required?

Question 2: What would be the cable size required using IEEE-45 approved cables?

<u>SOLUTION NO. 2</u>

A. First find the watts:

     $15 \times 746 = 11{,}190$

B. Next find the motor F.L.A.:

     $I = P \div E \times 0.8 \times \sqrt{3}$

     $I = 11190 \div 440 \times 0.8 \times 1.732$

     $I = 11190 \div 609.664$

     $I = 18.35$ F.L.A. (Round off to 18 Amps).

C. Size the circuit breaker (A) by multiplying the motor F.L.A. times 115%:

     $18 \times 1.15 = 20.70$ Amps (Round off to 21 Amps).

D. Using the enclosed Table A4-3, we see that the closest circuit breaker <u>Trip</u> size above 21 Amps is 25 Amps, therefore:

     Answer 1: The circuit breaker must be 100 Amp Frame / 25 Amp Trip.

E. The cable must carry at least 100% of the motor F.L.A. of 21, but not below the circuit breaker trip, which is 25 Amps.

F. Using the enclosed Table A4-4, pick the three conductor bronze armored cable that is the nearest to, but not below, 25 Amps with a rating of 75^{0} C.

     Answer 2: The cable must be a TTNB-10 which is rated at 32 Amps.

V.   MOTOR CALCULATIONS - PROBLEMS

In the power panel below fill in the motor F.L.A., the cable type and size, and the circuit breaker Frame and Trip size for each motor.

The first two circuits have already been filled in from Examples 1 and 2. Note that the circuit breaker Frame size precedes the Trip size, and both are separated by a slash mark.

440 VOLT-3 Ø
PWR. PNL P401

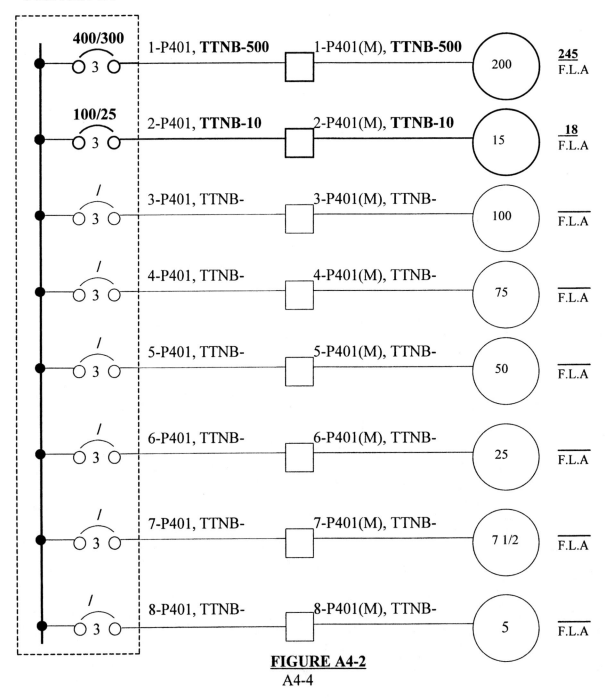

FIGURE A4-2
A4-4

TABLE A4-3 – Circuit Breaker Standard Sizes [from NEC 240-6]

AMP FRAME	AMP TRIP	AMP FRAME	AMP TRIP	AMP FRAME	AMP TRIP
100	15	100	90	400	400
100	20	100	100	1600	450
100	25	225	110	1600	500
100	30	225	125	1600	600
100	35	225	150	1800	700
100	40	225	175	1800	800
100	45	225	200	1800	1000
100	50	225	225	1800	1200
100	60	400	250	1800	1600
100	70	400	300		
100	80	400	350		

TABLE A4-4 – Distribution cables [from IEEE Std. 45-1998 Tables 8-33 and 9-1]

Three-conductor cables type TTN (A, B, or T) rated @ 45°C ambient temperature.
T = Three conductor distribution cable type   T = Polyvinyl Chloride Insulation
N = Thermosetting Polychloroprene (neoprene) Jacket
A = Aluminum armor or B = Bronze armor or T = Tin coated copper or None = No armor

MCM / AWG	CIRCULAR MIL	AMPS @ 75°C
14	4110	20
12	6530	24
10	10400	32
8	16500	41
7	20800	48
6	26300	54
5	33100	64
4	41700	70
3	52600	83
2	66400	93
1	83700	110
1/0	106000	126
2/0	133000	145
3/0	168000	168
4/0	212000	194
250 MCM	250000	217
300 MCM	300000	242
350 MCM	350000	265
400 MCM	400000	286
500 MCM	500000	329
535 MCM	535000	340
600 MCM	600000	368
750 MCM	750000	413

You have learned in lesson No. 4 – Motor Calculations, how to size each cable and circuit breaker from a power panel to a motor.

It is now necessary to size the feeder cable to the panel from its source of power, usually the Ships Service Switchboard.

It is also necessary to size the circuit breaker which protects this cable.

I.   Diagram of a typical Power Panel serving power to motors as shown on a One-line Diagram:

Lesson 5

440 VOLT-3 ∅
PWR. PNL P401

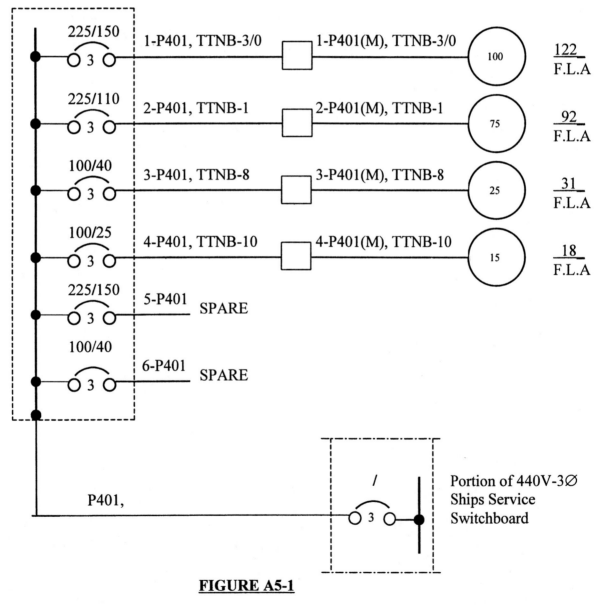

FIGURE A5-1

A5-1

II.   The U. S. Coast Guard requires that motor feeder cables be selected based on a computed demand load of the panel [U.S.C.G. 111.60-7 and table 111.60-7]. This table in turn references Article 430 of the National Electrical Code (NEC).

A. A cable supplying two or more motors must be rated for at least 125% of the full load amps of the largest motor, plus 100% for the sum of the full load amps for all the other motors [NEC 1999 Section 430-24]. In addition the connected load should include 50% of the trip ratings of the spare circuit breakers [IEEE Std.45-1998 Sub-clause 11.5]. This is the Demand Load Amps (D.L.A.) of the panel.

B. The circuit breaker must be rated by adding the the trip rating of the largest circuit breaker in the panel, plus the sum of the full load amps of the other motors connected to the panel [NEC 1999 Section 430-62(a)].

However, if the cable was rated to include spares for future additions, the circuit breaker could possibly be sized above the maximum rating of the cable, but cannot be over 150% of this rating [ABS Rule 4/5A5.1.3c].

The best way to rate this circuit breaker is to try to find the standard size trip rating which falls between the panel demand load and the rated amps of the cable.

III.   EXAMPLE NO. 1

Question 1: As shown in FIGURE A5-1 we have a six circuit 440 Volt – 3 phase power panel serving four motors and two spares. What would be the cable (A) size feeding this panel?

Question 2: What circuit breaker (B) Frame and Trip size will be required?

SOLUTION NO. 1:

A. First find the demand load on the panel:

1. Take 125% of the amp rating of the largest motor (100 HP); 122 F.L.A. x 1.25 = 152.50 Amps (round off to 153 Amps).

2. Add up the full load amps of the other motors on the panel 92 + 31 + 18 = 141 Amps.

3. Next add up half of the rating of the spare circuit breaker trips; 75 + 20 = 95 Amps.

4. Add the three totals from steps 1, 2, and 3; 153 + 141 + 95 = 389 Demand Load Amps (D.L.A.).

B. Using table A4-4 from Lesson 4, select a "TTNB" cable which carries <u>at</u> <u>least</u> 389 Amps.

   Answer 1: The cable must be a TTNB-750 which carries 413 Amps.

C. Adding up the largest circuit breaker trip (150) with the sum of the other three motor full load amps (92 + 31 + 18 = 141) we arrive at a required amps of 291, which does not allow any room for future growth. The best way to size the circuit breaker trips is to find the one that falls <u>between</u> the demand load amps of 389 and the cable amps of 413.

   Answer 2: As can be seen from Table A4-3 of Lesson 4, the nearest size circuit breaker trip which is above the demand load is a 400 Amp Trip.

   Therefore the circuit breaker should be a 400 Amp Frame – 400 Amp Trip.

IV. <u>EXAMPLE NO. 2</u>

   Question 1: In the power panel shown in Figure A5-3, what are the circuit breaker and cable sizes required for each individual motor?

   Question 2: What size feeder cable and circuit breaker are required?

   <u>SOLUTION NO. 2</u>:

A. In Lesson 4 we learned to calculate the amp rating of each motor through the use of a formula and using 1 Horsepower = 746 Watts.

   This formula was taught to you for unusually large motors, and is not necessary for this lesson. Most motor amps can be found in standard tables, thus eliminating the need for extra mathematics.

   From Table A5-2 [NEC 1999 Table 430-150] plug in the full load amps of each motor in Figure A5-3 using the 440/480 Volt column. The assumption will be made that all motors are the induction type.

B. Proceed with sizing each motor cable and circuit breaker as learned in Lesson 4. Plug these sizes into Figure A5-3.

   You will notice that the current carrying capacity of the cables (ampacity) is usually greater than the circuit breaker trip size; thus the circuit breaker protects the cable which supplies the motor. The motor is protected by overloads in the controller (C).

C. Size the spare circuit breakers in the panel.

1. The quantity of spares in a power panel is determined by the Owners specification requirements, generally one spare for every five active circuits or portions thereof. For the purpose of this lesson the quantity of spares is shown.

2. Spares are usually figured at one spare for each <u>Frame</u> size used. We have used 225 Amp Frame and 100 Amp Frame active circuits, therefore we should have one Frame size for each as spares.

3. The trip rating for each Frame size should be either the trip size most commonly used, or an average of the active Trip sizes for that Frame rating.

4. The most used trip at 225 A.F. is 150 – therefore the 225 A.F. spare should have a 150 A.T.

5. The most used trip at 100 A.F. is 15 – therefore the 100 A.F. spare should have a 15 A.T.

D. Find the Demand Load Amps (D.L.A.) for the panel and plug this value into the diagram.

1. Take 125% of the amp rating of the largest motor (100 H.P.): 124 x 1.25 = 155.

    NOTE: When two motors are rated the same, only use one at 125%, and the other at 100% of F.L.A.

2. Add up the full load amps of the other motors on the panel: 124 + 96 + 65 + 34 + 40 + 11 + 7.6 = 377.6

3. Add up half of the rating of the spare circuit breaker trips: 75 + 7.5 = 82.5

4. Add up the amps arrived at in steps 1, 2, and 3 above: 155 + 377.6 + 82.5 = 615.10 <u>Demand Load Amps</u> (Round off to 615 D.L.A.).

E. Using Table A4-4 from Lesson 4, select a "TTNB" cable that carries <u>at</u> <u>least</u> 615 Amps.

NOTE: If the demand load is greater than one cable is rated, split the demand load and use two cables of the same size, each rated for at least half the demand load, which is 307.5 Amps.

Answer 1: The cables must be two (2) TTNB-500's, which each carry 329 Amps for a total of 658 Amps.

F. Using Table A4-3 from Lesson 4, select a circuit breaker Trip size which falls between the demand load amps of 615 and the cable combined amps of 658.

Answer 2: As can be seen from the table, the nearest size circuit breaker trip which is above the demand load is a 700 Amp trip.

Even though this rating is above the cable rating, it does not exceed 150% of the cable rating (658 x 150% = 987 Amps).

Therefore the circuit breaker must be an 1800 Amp Frame – 700 Amp Trip.

TABLE A5-2 – Full-Load Current Three Phase Alternating-Current Motors
(from NEC Table 430-150)

Horsepower	Induction Type Squirrel Cage and Wound Rotor (Amperes)					Synchronous Type Unity Power Factor* (Amperes)		
	110/120 Volts	208 Volts	220/240 Volts	440/480 Volts	550/600 Volts	220/240 Volts	440/480 Volts	550/600 Volts
½	4.4	2.4	2.2	1.1	0.9	-	-	-
¾	6.4	3.5	3.2	1.6	1.3	-	-	-
1	8.4	4.6	4.2	2.1	1.7	-	-	-
1½	12.0	6.6	6.0	3.0	2.4	-	-	-
2	13.6	7..5	6.8	3.4	2.7	-	-	-
3	-	10.6	9.6	4.8	3.9	-	-	-
5	-	16.7	15.2	7.6	6.1	-	-	-
7½	-	24.2	22	11	9	-	-	-
10	-	30.8	28	14	11	-	-	-
15	-	46.2	42	21	17	-	-	-
20	-	59.4	54	27	22	-	-	-
25	-	74.8	68	34	27	53	26	21
30	-	88	80	40	32	63	32	26
40	-	114	104	52	41	83	41	33
50	-	143	130	65	52	104	52	42
60	-	169	154	77	62	123	61	49
75	-	211	192	96	77	155	78	62
100	-	273	248	124	99	202	101	81
125	-	343	312	156	125	253	126	101
150	-	396	360	180	144	302	151	121
200	-	528	480	240	192	400	201	161
250	-	-	-	302	242	-	-	-
300	-	-	-	361	289	-	-	-
350	-	-	-	414	336	-	-	-
400	-	-	-	477	382	-	-	-
450	-	-	-	515	412	-	-	-
500	-	-	-	590	472	-	-	-

*For 90 and 80 percent power factor, the figures shall be multiplied by 1.1 and 1.25 respectively.

440 VOLT-3 ∅
PWR. PNL P402

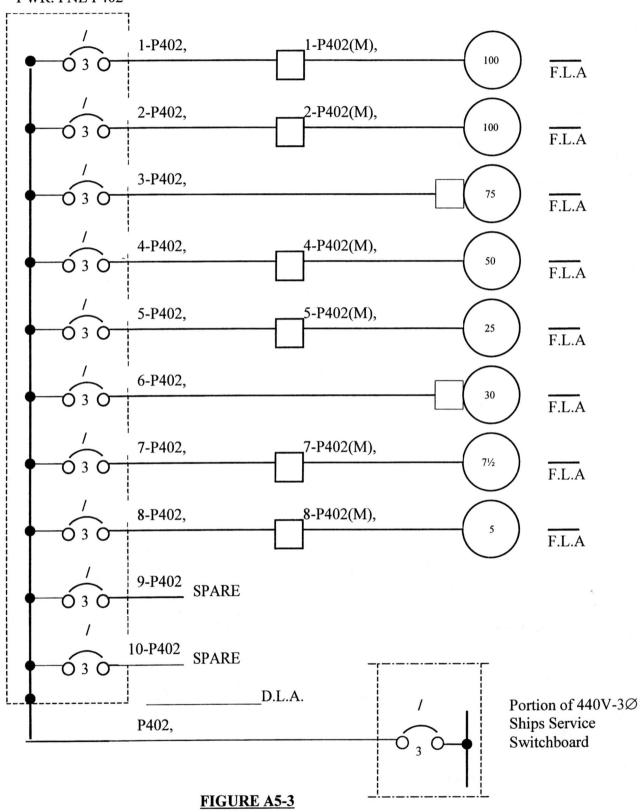

FIGURE A5-3

V.   POWER PANEL FEEDERS – PROBLEM NO. 1

A. In the panel below, plug in the F.L.A. of each motor from Table A5-2, select each TTNB cable and circuit breaker, and size the feeder cable and breaker.

440 VOLT-3 Ø
PWR. PNL P403

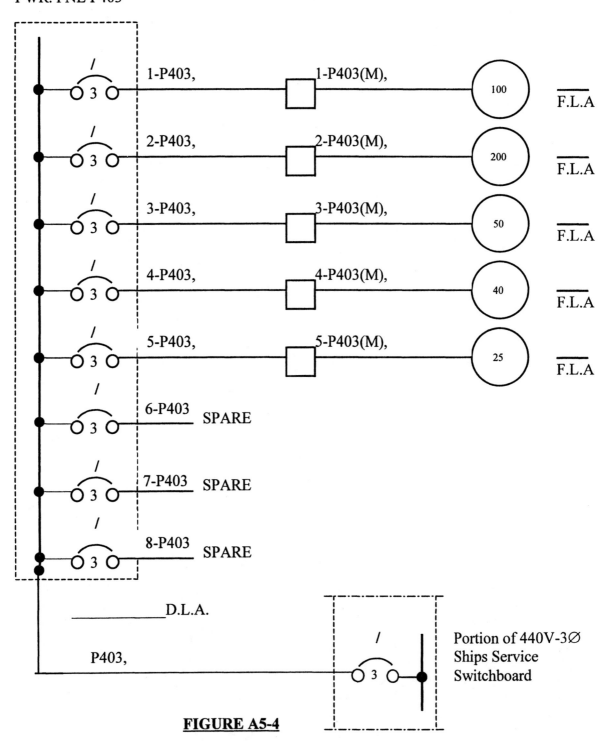

<u>FIGURE A5-4</u>

VI.   POWER PANEL FEEDERS – PROBLEM NO. 2

   A. In the power panel below, size the cables, breakers, and feeder sizes. Also plug
       in the appropriate circuit numbers.

440 VOLT-3 Ø
PWR. PNL P404

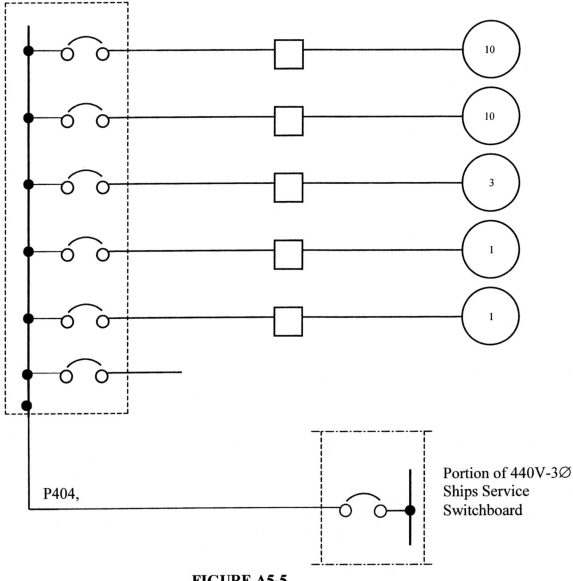

P404,

Portion of 440V-3Ø
Ships Service
Switchboard

FIGURE A5-5

VII.   POWER PANEL FEEDERS – PROBLEM NO. 3

A. Size the cables, breakers, and feeder in the diagram below. Plug in circuit numbers.

440 VOLT-3 ∅
PWR. PNL P405

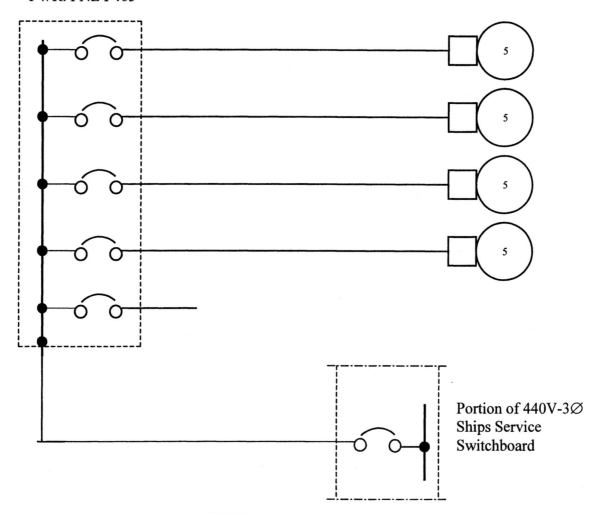

FIGURE A5-6

VIII.   POWER PANEL FEEDERS – PROBLEM NO. 4

A. On a separate sheet, draw a One-Line Diagram of a power system which includes Examples 1 and 2, and problems 1, 2, and 3 of this lesson. Put in all motors, cables, breakers, and circuit designations all originating from the single Ship's Service Switchboard.

MARINE ELECTRICAL BASICS

WORKBOOK

Module B

INTERMEDIATE

You have already learned in Lessons 4 and 5 how to size the cables and circuit breakers for motors and power panels. It is also necessary to size lighting panels and small appliance panels in accordance with United States Coast Guard regulations and ABS Rules.

I.   Diagram of a typical Lighting Panel as shown on a One-Line Diagram:

120 VOLT-3Ø MN., 1Ø BRCHS.
LTG. PNL L101   **FIGURE B6-1**

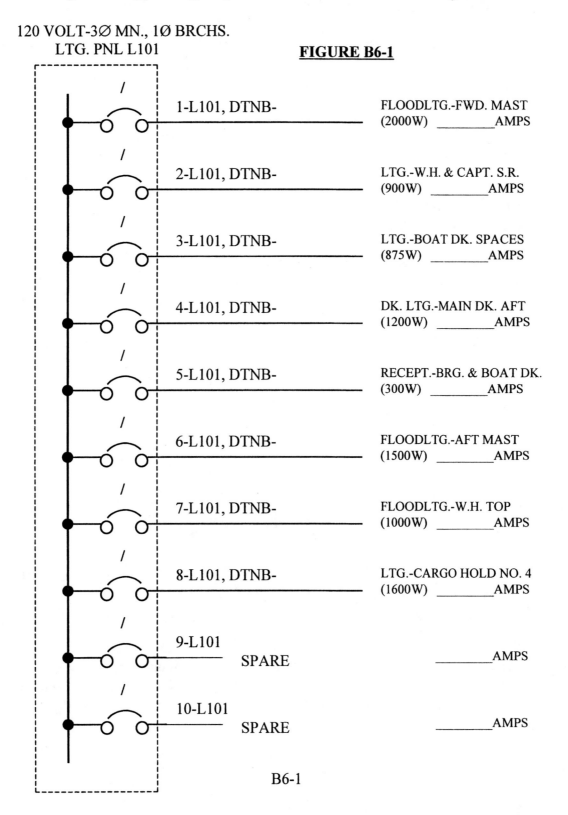

1-L101, DTNB-	FLOODLTG.-FWD. MAST (2000W) _____AMPS
2-L101, DTNB-	LTG.-W.H. & CAPT. S.R. (900W) _____AMPS
3-L101, DTNB-	LTG.-BOAT DK. SPACES (875W) _____AMPS
4-L101, DTNB-	DK. LTG.-MAIN DK. AFT (1200W) _____AMPS
5-L101, DTNB-	RECEPT.-BRG. & BOAT DK. (300W) _____AMPS
6-L101, DTNB-	FLOODLTG.-AFT MAST (1500W) _____AMPS
7-L101, DTNB-	FLOODLTG.-W.H. TOP (1000W) _____AMPS
8-L101, DTNB-	LTG.-CARGO HOLD NO. 4 (1600W) _____AMPS
9-L101 SPARE	_____AMPS
10-L101 SPARE	_____AMPS

Lesson 6

II.   U.S. Coast Guard and IEEE-45 rules for sizing cables and circuit breakers for lighting and small appliance circuits are:

    A. Circuit breakers or switches in grounded neutral distribution panels (such as three phase main lighting panels) should include a pole for the neutral [IEEE Std. 45-1998 Sub-clause 13.1]. This requires all circuit breakers in lighting panels to be two pole. If the fixtures are located in a hazardous area, the switches or protective devices <u>must </u>interrupt all poles or phases, and be located in a non-hazardous area [ABS Rule 4/5B7.1.2].

    B. The same IEEE Std. 45-1998 Sub-clause 13.1 limits the number of branch circuits controlled by a distribution panel to a maximum of 18 for 3 phase, or 26 for single phase. This limits the number of two pole circuit breakers in a lighting panel to 26, or a panel having no more than 52 overcurrent devices (poles).

    C. A lighting distribution panel cannot supply circuits rated over 30 amperes [U.S.C.G. Subpart 111.75-5(a)]. Overload protective devices cannot be rated over 30 amperes [ABS Rule 4/5A7.1.4].

    D. The connected load on any normal lighting circuit cannot exceed 80% of its circuit breaker rating [U.S.C.G. Subpart 111.75-5(b) and ABS Rule 4/5A7.1.4]. The circuit breaker must be set at 15 or 20 Amp Trip [U.S.C.G. Subpart 111.75-5(d)]. The cable cannot be smaller than 14 AWG ($2.10mm^2$) [U.S.C.G. Subpart 111.60-4(a)].

    E. The connected load on most fixed nonswitched deck or cargo hold lighting circuits with lamp holders exceeding 300 watts (usually floodlighting), cannot exceed 24 amps. The circuit breaker must be set at 25 or 30 amps and the cable cannot be smaller than 10 AWG ($5.3mm^2$) [U.S.C.G. 111.75-5(e)].

III.   <u>EXAMPLE NO. 1:</u>

Question 1: As shown in Figure B6-1, we have a ten circuit 120 Volt 3 Phase Main – 1 Phase Branches Lighting Panel serving eight active circuits with two spares. What would be the circuit breaker size for each circuit?

Question 2: Using "DTNB" type cables as shown in Table B6-6, what would be the cable size for each circuit?

<u>SOLUTION NO. 1:</u>

    A. It is standard practice in the marine field to have all circuit breakers in lighting panels rated at 100 Amp Frame. Therefore plug in this figure (100) to the left of the slash mark for all ten circuit breakers.

B. Since all the branch circuits are single phase, each cable must be a two conductor, or "DTNB" type, and each circuit breaker must have a pole for each conductor. This means each circuit breaker must be a two pole. Plug this figure (2) into each circuit breaker symbol beneath the curved line.

There will never be more than 26 circuits on a lighting panel since IEEE Std. 45-1998 does not allow more than 52 overcurrent devices (poles) on a lighting panel.

C. Next find the connected load of each active circuit in amps. This is done by the formula: I = P ÷ E; where I = current in amps, P = power in watts, and E = electromotive force in volts.

The voltage in the case of 3 phase – 120 Volt or 208Y/120 Volt lighting panels in most cases on shipboard is 120 Volts for each single phase lighting branch circuit. This is true whether the transformers or supply feeding the panel is connected delta-delta, or delta-wye.

By utilizing this formula for circuit 1-L101 in figure B6-1, we see that the connected load equals 2000 watts divided by 120 volts, or <u>16.67</u> amps.

Calculate the amps for each circuit and plug these figures into the diagram.

NOTE: For the purpose of this lesson, the connected load for any of the circuits shown does not exceed that allowed by U.S.C.G. and A.B.S. regulations. It is the responsibility of the person designing or hooking up the lighting system not to exceed the amps on any one circuit from what is allowed. For example, if a 15 Amp Trip lighting circuit has more than 1380 watts (12 amps) connected, this circuit must be split into two separate circuits, or the circuit breaker increased to 20 Amps.

D. Now that the connected load in amps has been calculated, the trip rating and cable size for each circuit in the panel can be determined.

1. Any circuit with 12 Amps or less will have a 15 Amp Trip and a 14 AWG (American Wire Gauge) cable. This may include any deck lighting or cargo hold lighting circuits which may be less than 12 Amps and are not floodlights. Plug this figure (15) to the right of each appropriate circuit breaker.

Per the enclosed Table B6-6, we see that the 14 AWG cable size is a DTNB-14 which is rated at 24 Amps. Plug this into Figure B6-1 where each 15 Amp Trip circuit breaker is used.

2. Any circuits between 12.1 and 16 Amps will have a 20 Amp Trip and use 14 AWG cable. This may also include deck lighting or cargo hold lighting which is not floodlighting. Plug this figure (20) into the circuit breakers as appropriate.

    The cable, per the enclosed Table B6-6, will be a "DTNB-14" also.

3. Any flood lighting circuits 20 amps and under will have a 25 Amp Trip, and any between 20.1 and 24 amps will have a 30 Amp Trip. All floodlight circuits must use at least a 10 AWG cable.

    Plug this figure (25) into the diagram for the floodlight circuits. Using Table B6-6, we find that a 10 AWG cable is a "DTNB-10". Plug this into the diagram.

E. Size the spare circuit breakers in the panel:

1. The quantity of spares in a lighting or appliance branch circuit panel board is usually determined by the owner's specification requirements. This is generally one spare circuit available for every ten active circuits, or portions thereof. For the purpose of this lesson, the quantity of spares is shown.

2. The trip rating of spares is arrived at by figuring the average load of all the active circuits ($16.67 + 7.50 + 7.29 + 10.00 + 2.50 + 12.50 + 8.33 + 13.33 = 78.12$), and dividing by the number of active circuits ($78.12 \div 8 = 9.77$ Amps).

3. It can be seen that the 9.77 amps is less than 12 amps, therefore the trip rating of each spare is 15 amps. Plug this figure (15) into the diagram.

IV.   EXAMPLE NO. 2:

Question 1: In the Appliance Panelboard in Figure B6-2, what are the circuit breaker and cable sizes required?

120 VOLT-3∅ MN., 1∅ BRCHS.
GALLEY PWR. PNL P102

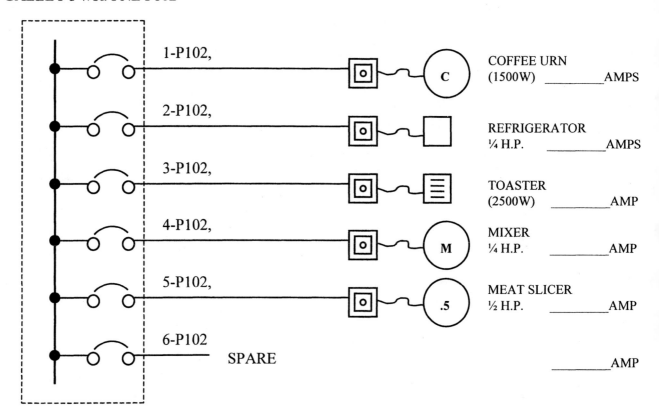

FIGURE B6-2

SOLUTION 1:

A. The same rules of connected loads and cables which apply to lighting panels also apply to small single phase appliance panels:

1. This means that a load of;

12 amps or less must have a 15 Amp Trip and a 14 AWG cable,
12.1 to 16 amps must have a 20 Amp Trip and a 14 AWG cable
16.1 to 20 amps must have a 25 Amp Trip and a 12 AWG cable
20.1 to 24 amps must have a 30 Amp Trip and a 10 AWG cable

2. Branch circuits for fixed appliances, or receptacles for special applications, should be on separate branch circuits [IEEE Std. 45-1998 Sub-clause 11.23 and 11.24]. This means receptacles which are designated for specific equipment should be on separate branch circuits.

B. Using the enclosed Table B6-7 we can find the amps of each small appliance that has a motor. Plug this figure into the diagram beside the refrigerator, mixer, and slicer.

C. Find the amps of the toaster and coffee urn by the formula $I = P \div E$. Plug this figure into the diagram. Remember that the branch voltage is 120 Volts.

D. Now that we have the amp load of each circuit we can figure the breaker trip size and cable rating for each circuit:

1. It is already known that each frame size will be 100, and each breaker will have 2 poles. Since circuits 2, 4, and 5 have each less than 12 amps, the trip rating must be 15, and the cable size must be a DTNB-14.

2. Circuit 1-P102 has an amp rating which falls between 12.1 and 16, therefore the trip must be 20 and the cable must be a DTNB-14.

3. Circuit 3-P102 has an amp rating between 20.1 and 24, therefore the trip must be 30, and the cable must be a DTNB-10.

E. Size the spare circuit breaker in the panel. The quantity of spares in a lighting or appliance branch circuit panel board is usually determined by the owners specification requirements, generally one spare for every ten active circuits or portions thereof. For the purpose of this lesson, the quantity of spares is shown.

1. Total up the loads ($12.50 + 5.8 + 20.83 + 5.8 + 9.8 = 54.73$), and divide by the active circuits ($54.73 \div 5 = 10.95$ amps). Since this is below 12 amps, the trip rating of the spare circuit breaker must be 15 amps.

V.   LIGHTING CALCULATIONS – PROBLEM NO. 1
        A. In the lighting panel in Figure B6-3, plug in the connected load in amps, and size
            each circuit breaker and cable. Use "DTNB" cables as listed in Table B6-6.

VI.   LIGHTING CALCULATIONS – PROBLEM NO. 2

        A. In the lighting panel in Figure B6-4, plug in the circuit numbers, connected load
            in amps, and cable and breaker sizes.

VII.   LIGHTING CALCULATIONS – PROBLEM NO. 3

        A. In the appliance panel in figure B6-5, plug in the connected load in amps, circuit
            numbers, cable and breaker sizes.

120 VOLT-3Ø MN., 1Ø BRCHS.
LTG. PNL L103

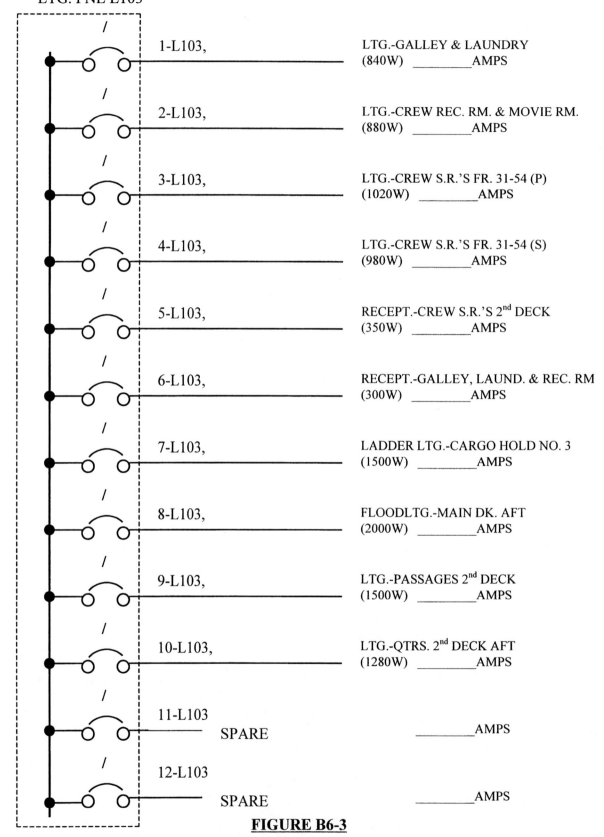

FIGURE B6-3

120 VOLT-3∅ MN., 1∅ BRCHS.
LTG. PNL L104

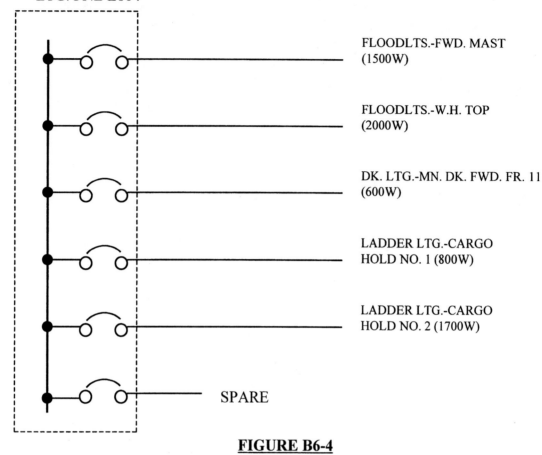

FLOODLTS.-FWD. MAST
(1500W)

FLOODLTS.-W.H. TOP
(2000W)

DK. LTG.-MN. DK. FWD. FR. 11
(600W)

LADDER LTG.-CARGO
HOLD NO. 1 (800W)

LADDER LTG.-CARGO
HOLD NO. 2 (1700W)

SPARE

FIGURE B6-4

120 VOLT-3∅ MN., 1∅ BRCHS.
GALLEY PWR. PNL P105

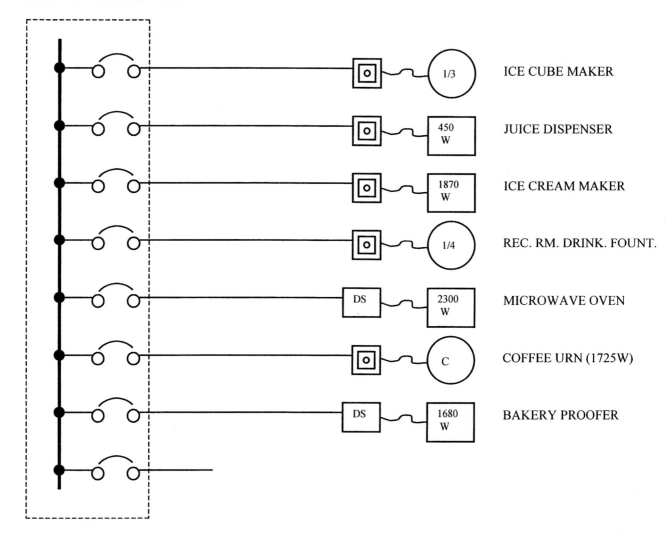

ICE CUBE MAKER

JUICE DISPENSER

ICE CREAM MAKER

REC. RM. DRINK. FOUNT.

MICROWAVE OVEN

COFFEE URN (1725W)

BAKERY PROOFER

FIGURE B6-5

TABLE B6-6 – Distribution cables[from IEEE Std. 45-1998 Tables 8-33 and 9-1]

Two-conductor cables type DTN (A, B, or T) rated @ 45⁰C ambient temperature.
D = Two conductor distribution cable type   T = Polyvinyl Chloride Insulation
N = Thermosetting Polychloroprene (neoprene) Jacket
A = Aluminum armor or B = Bronze armor or T = Tin coated copper or None = No armor

MCM / AWG	CIRCULAR MIL	AMPS @ 75⁰C
14	4110	24
12	6530	31
10	10400	38
8	16500	49
7	20800	59
6	26300	66
5	33100	78
4	41700	84
3	52600	102
2	66400	115
1	83700	134
1/0	106000	153
2/0	133000	187
3/0	168000	205
4/0	212000	237
250 MCM	250000	264

TABLE B6-7 – Full-Load Current Single Phase Alternating Current Motors
(from NEC Table 430-148)

Horsepower	115 Volts	200 Volts	208 Volts	230 Volts
1/6	4.4	2.5	2.4	2.2
1/4	5.8	3.3	3.2	2.9
1/3	7.2	4.1	4.0	3.6
1/2	9.8	5.6	5.4	4.9
3/4	13.8	7.9	7.6	6.9
1	16	9.2	8.8	8.0
1-1/2	20	11.5	11.0	10
2	24	13.8	13.2	12
3	34	19.6	18.7	17
5	56	32.2	30.8	28
7-1/2	80	46.0	44.0	40
10	100	57.5	55.0	50

Voltages listed are rated motor voltages. The currents listed shall be permitted for system voltage ranges of 110 to 120 and 220 to 240 volts.

You learned, in Lesson No. 6 – Lighting Calculations, how to size each circuit breaker and cable from a Lighting Panel. It is now necessary to size the feeder cable to the panel from its source of power, usually the 120 Volt – 3 phase section of the Ships Service Switchboard.

It is also necessary to size the circuit breaker which protects this cable.

I.   Diagram of a typical Lighting Panel as shown on a One-Line Diagram:

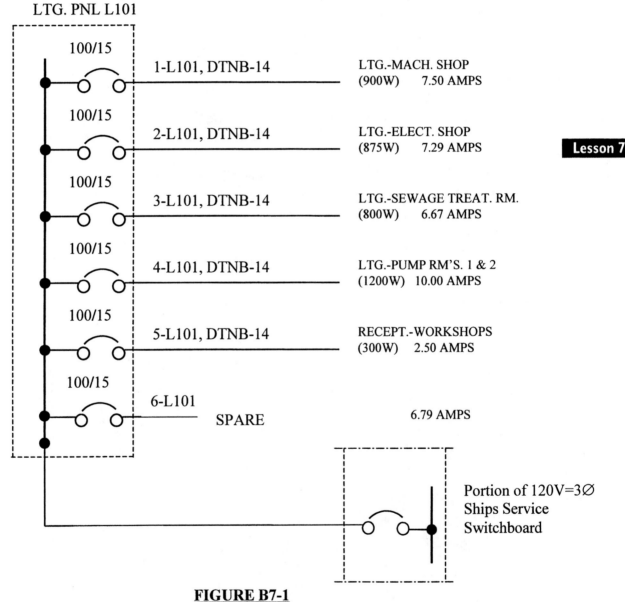

FIGURE B7-1

II.   U. S. Coast Guard and ABS Rules for sizing cables and circuit breakers feeding lighting panels:

    A. A cable supplying a lighting panel shall be rated at least 100% of the connected load, plus the average of the active load for each spare circuit on the panel. This is the <u>demand load</u> on the panel [U.S.C.G. Table 111.60-7].

    B. The circuit breaker is designed to protect the cable from overheating, and should not be rated any larger than the current carrying capacity of the conductor. However, if the capacity of the cable does not correspond to a standard circuit breaker trip rating, the next higher breaker trip may be used, if it does not exceed 150% of the cable ampacity [U.S.C.G. Sub-part 111.50-3(c) and ABS Rule 4/5A5.1.3c].

III.   <u>EXAMPLE NO. 1</u>:

Question 1: As shown in Figure B7-1, we have a six circuit lighting panel with five active circuits and one spare. What would be the cable size feeding this panel?

Question 2: What would be the circuit breaker size required in the Ships Service Switchboard?

<u>SOLUTION NO. 1</u>:

    A. First find the demand load on the panel:

        1. You may think that since the amps for each branch circuit has already been calculated, the demand load can be arrived at by simply adding up the branch circuit amps listed. <u>This is not the case</u> since the branch circuit is single phase 120 volts, and the feeder voltage to the panel is three phase 120 volts.

        2. To find the correct demand load on the panel we must first add up the watts, not amps, for each circuit (900 + 875 + 800 + 1200 + 300 = 4075 Watts).

        3. Next we must calculate the watts for each spare by the formula:
$$P = E \times I, \quad P = 120 \times 6.79, \quad P = 815 \text{ watts.}$$

        4. Since our panel only has one spare circuit, add the spare watts to the connected load (815 + 4075 = 4890 watts).

        5. Convert this total wattage to amps by the following formula for three phase circuits:

$$I = P \div E \times \sqrt{3}$$

$$I = 4890 \div 120 \times 1.732$$

$$I = 4890 \div 207.84$$

$$I = 23.53 \text{ Demand Load Amps (Round off to 24 Amps).}$$

B. Now that the demand load on the panel is known, the feeder cable type and size can be selected:

1. Since the voltage for the panel <u>main</u> is three phase, we know that the cable must be a three conductor.

2. Using the "TTNB" cable chart (Table A4-4 of Lesson No. 4) we see that the cable can be a TTNB-12 which has an ampacity of exactly 24 amps.

3. Although this arrangement would satisfy the USCG and ABS Rules, it is not a good idea to tie ourselves so closely to minimum allowable values. There is a very good reason why we should increase the feeder cable size.

   a. As the designer or installer will learn in later lessons, the voltage drop for cables on 120 volt circuits adds up so rapidly per foot, that any cable figured this close would eventually be changed to the next larger size.

4. Therefore, in accordance with <u>good</u> <u>marine</u> <u>practice</u>, whenever we have a circuit breaker feeding a lighting or appliance branch panel which has a trip setting so close to the panel demand load, we should <u>always</u> go to the next larger cable. In this case we would use a TTNB-10 cable which carries 32 amps. Plug this size into Figure B7-1.

C. Pick a circuit breaker trip rating which is between the panel demand load (24 amps) and the cable amps (32 amps).

1. Using Table A4-3 of Lesson 4 we see that the circuit breaker will be rated 100 Amp Frame / 30 Amp Trip. Plug this figure into the diagram at the main switchboard.

2. We already know that the number of poles will be three since the panel has a three phase main. Plug this figure (3) under the curve of the circuit breaker (B).

IV. <u>EXAMPLE NO. 2:</u>

Question 1: In the appliance panel shown in Figure B7-2, what are the circuit breaker and cable sizes for each branch circuit?

Question 2: What size feeder cable and circuit breaker is required?

<u>SOLUTION NO. 2:</u>

A. First find the amp rating of the toaster and coffee urn by the formula I = P ÷ E. Remember the branch circuits are rated at 120 volts.

B. Next find the amp ratings of the motors from Table B6-7 of Lesson No. 6.

C. Calculate the average amp rating for spare circuits and plug this value into the diagram for each spare.

D. Size all branch circuit breakers and cables in the panel as shown in lesson No. 6.

E. Find the demand load on the panel:

1. First convert the amps to watts for each branch circuit by the formula P = E x I. This has already been done for the toaster and coffee urn.

2. Add up the watts for each circuit including spares. This should total <u>9044</u> watts.

3. Convert this wattage to three phase amps at 120 volts per the formula:

$$I = P \div E \times \sqrt{3}$$

$$I = 9044 \div (120 \times 1.732)$$

$$I = 9044 \div 207.84$$

$$I = 43.51 \text{ Demand Load Amps (Round off to 44 Amps).}$$

F. Using the "TTNB" cable chart (Table A4-4 from Lesson No. 4), we see that a TTNB-7 carries 48 amps at 75^0 rated temperature.

G. Using Table A4-3 of Lesson 4, pick a circuit breaker rating which is between the panel demand load of 44 amps and the cable amps (48 amps). The circuit breaker will be rated 100 Amp Frame / 45 Amp Trip. Plug the Frame, Trip, and poles (3) into Figure B7-2.

1. Again this arrangement would satisfy the USCG and ABS Rules but we are tying ourselves too closely to minimum allowable values.

2. Therefore, again in accordance with <u>good marine practice</u>, we should go to the next larger cable. In this case we would use a TTNB-6 cable which carries 54 amps. Plug this size into Figure B7-2.

120 VOLT-3Ø MN., 1Ø BRCHS.
MESS RM. PWR. PNL P102

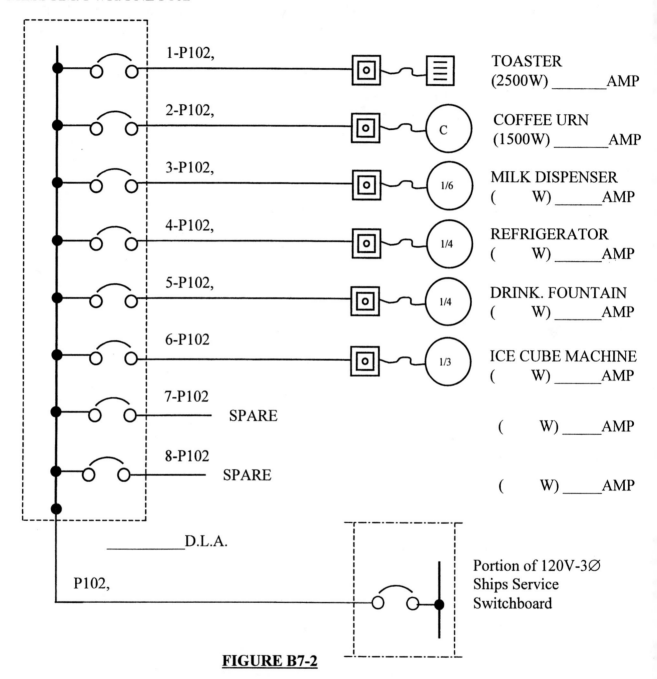

TOASTER
(2500W) _____AMP

COFFEE URN
(1500W) _____AMP

MILK DISPENSER
(W) _____AMP

REFRIGERATOR
(W) _____AMP

DRINK. FOUNTAIN
(W) _____AMP

ICE CUBE MACHINE
(W) _____AMP

(W) _____AMP

(W) _____AMP

_____D.L.A.

P102,

Portion of 120V-3Ø
Ships Service
Switchboard

FIGURE B7-2

V.   LIGHTING PANEL FEEDERS – PROBLEM NO. 1

    A. In problems 1, 2, and 3 of Lesson No. 6, you sized the branch cables and circuit breakers for lighting and appliance panels L103, L104, and P105.

    B. On a separate sheet, draw a One-Line diagram which includes examples 1 and 2 of <u>this</u> lesson, and problems 1, 2, and 3 of Lesson No. 6, showing the feeder cables and circuit breakers from the Ships Service Switchboard.

    C. Figure the Demand Load Amps, then add the size of the feeder cables and circuit breakers on this One-Line for panels L103, L104, and P105 from Lesson 6.

Basic ships power is usually rated at 440 volts–3 phase–60 cycles (Hertz), derived from Ships' Service Switchboards. See Table B8-6 of this lesson for standard ships' voltages.

This must be converted, or transformed, into various different voltages for ships' lighting, and other lower voltage services.

This is generally accomplished by the use of transformers, which are electrical devices designed to convert alternating current from one voltage to another. Transformers can be designed to either "step-up" or "step-down" voltages.

I.   Diagram of a Ships' Service main lighting transformer installation as shown on a One-Line Diagram:

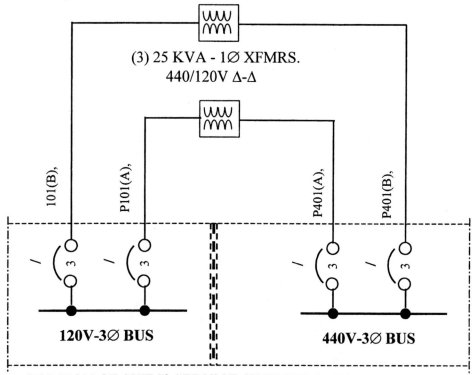

PORTION OF SHIPS' SERVICE SWITCHBOARD

FIGURE B8-1

II.   The rules for sizing transformer cables and overcurrent protections are:

A. Transformers must have overcurrent protection which meets Article 450 of the NEC or IEC 92-303 [U.S.C.G Sub-part 111.20-15].

B. Transformers of 600 Volts or less must have primary overcurrent protection which does not exceed 125% of the rated primary current of the transformer [NEC 1999 Article 450-3(b) with Table 450-3(b) and ABS Rule 4/5A5.15.1]. To achieve the best design and installation, the protective device should be rated for 100% of the transformer rating even though the transformer may not be fully loaded.

C. Transformers which are an essential part of the propulsion or ships service electrical supply must be duplicated similar to the requirement for generators so as to supply power in the event of failure without using the emergency system [USCG Sub-part 111.10-9 and ABS Rules 4/5A4.1.6a and 4/5A2.1.2]. The main and stand-by transformers are usually connected in parallel as shown in Figure B8-1.

D. When transformers are arranged for parallel operation, means must be provided to disconnect the secondary circuits via an isolating switch and short circuit protection, or a circuit breaker[ABS Rule 4/5A4.1.6b, Rule 4/5A5.15.2, and IEEE Std. 45-1998 Sub-clause 34.5].

   1. If the transformer has an overcurrent device in the secondary circuit is must be rated at no more than 125% of the secondary current, but the primary circuit protection may be set at a value less than 250% of the primary current –ABS Rule 4/5A5.15.1 and NEC Table 450-3(b)].

   2. This secondary circuit breaker is not required on single phase transformers, or three phase delta-delta connected transformers which have a three wire secondary cable [USCG Sub-part 111.50-5(a)(4) and NEC Article 240-3(f).

E. The neutral conductors of some 3 phase, 4 wire wye circuits are considered to be current carrying conductors due to harmonic currents which may be present [NEC Article 310-15(b)(4)(c)].

   1. Although not specifically required by the rules, due to the ability to easily overload circuits which are fed by a three phase four wire delta-wye transformer with a neutral, it is recommended that any transformers with dual voltage delta-wye secondaries be protected by a four pole overcurrent device in the secondary circuit. This would apply to transformers feeding some galley or laundry panels utilizing 120-208V delta-wye secondaries as shown in Figure B8-4 and Figure B8-5.

F. Although the rules do not have specific instructions for sizing transformer cables, we can deduce that the cable must be rated at a higher ampacity than the circuit breaker trip, since the breaker is designed to protect the cable.

   1. Sometimes cables which have more than three conductors must be "derated". [U.S.C.G. Sub-part 111.60-3(c) and NEC Table 310-15(b)(2)(a)]. That is, a percentage must be applied to the listed ampacity for three conductor cables, unless the manufacturer has a chart listing the ampacities of four conductor cables. When a four conductor cable is used, such as in delta-wye secondary circuits where the fourth conductor is a neutral, the ampacity does not need to be adjusted from that listed for a three conductor cable [IEEE Std. 45-1999 Table 9.1 note 3].

III.   EXAMPLE NO. 1

Question 1: As shown in Figure B8-1 we have a parallel set of three (3) single phase 25 KVA transformers with a primary voltage of 440 and a secondary voltage of 120. Each bank of transformers is connected together in a Delta-Delta (Δ - Δ) configuration to make a three-phase transformer bank. What would be the circuit breaker sizes required for the primary cables and secondary cables?

Question 2: What cable sizes are required for the primary and secondary feeders?

SOLUTION NO. 1:

A.   The first thing to do is figure the total power of each transformer bank. Since both banks are identical it is only necessary to calculate this once.

  1. Three phase transformers are sometimes not readily available whereas single phase transformers are generally found in stock.

  2. Three (3) single phase transformers, of equal size and characteristics, can be connected together to form a three phase transformer bank.

  3. The equivalent three phase capacity, when properly connected, is three times the nameplate rating of one single phase transformer.

  4. In this case, three 25 KVA single phase transformers will accommodate a 75 KVA (3 x 25 KVA) three phase load.

B.   The next step is to determine the current of the primary and secondary sides of the transformer bank.

  1. A transformer is a completely static solid state device with no moving parts, which works on the magnetic induction principal. It consists of two or more coils of wire wound on a laminated steel core.

  2. When voltage is introduced to one coil, called the primary coil, it magnetizes the iron core. A voltage is then induced into the other coil, called the output or secondary coil. The change of voltage between the primary and secondary depends on the turns ratio of the two coils.

  3. In this example, we have a primary voltage of 440 and a secondary voltage of 120.

  4. To find the current we use the following formula for three phase circuits (see Table B8-7 of this lesson):

$$I = KVA \times 1000 \div 1.732 \times E$$

5. Since we know the primary voltage is 440 and the total power is 75 KVA, we find the primary current by:

    $I = 75 \times 1000 \div 1.732 \times 440$
    $I = 75000 \div 762.08$
    $I = 98.41$ amps primary current (round off to 98 amps).

6. Secondary current is found by the same formula but using 120 volts:

    $I = 75 \times 1000 \div 1.732 \times 120$
    $I = 75000 \div 207.84$
    $I = 360.85$ amps secondary current (round off to 361 amps).

C. Now that we know the amp rating of the primary and secondary sides it is necessary to size the circuit breakers.

1. The circuit breaker trip settings should be rated at 100% of the currents but never over 125% of this current.

    Primary is $98 \times 1.25 = 122.50$ amps (round off to 123 amps)
    Secondary is $361 \times 1.25 = 451.25$ amps (round off to 451 amps)

2. We can go back to Table A4-3 of Lesson 4 and select a circuit breaker with a trip rating between 98 and 123 amps. Plug this figure 100AF / 100 AT into the diagram for both primary circuit breakers (A).

3. We can go use this same Table A4-3 and select a circuit breaker with a trip rating between 361 and 451 amps. Plug this figure 400AF / 400 AT into the diagram for both secondary circuit breakers (B).

4. We also know the circuit breakers for the primary must be three pole since the circuits are three phase. Make sure this figure (3) is plugged into the breaker symbols for P401(A) and P401(B).

5. The circuit breakers for the secondary circuits must be three pole since the circuits are three phase. Make sure this figure (3) is plugged into the breaker symbols for P101(A) and P401(B).

D. The last thing we have to do is size the cables for the primary and secondary sides of the transformer:

1. The primary cables must be rated at least 100 amps (the circuit breaker trip setting. We also know they must be three conductors rated for an ambient temperature of 45^0C.

2. Using Table A4-4 of Lesson 4, we see that the cable closest to, but not below, 100 amps is a TTNB-1 which has an ampacity of 110 amps at 45^0C. Since this does not exceed the 125% rating of 123 amps, the selected cable is correct. This cable can be plugged into the diagram for the primary cables P401(A) and P401(B).

3. Single phase transformers are usually connected into three phase "banks" by one of two means, depending on the voltage requirements of the vessel. The primary side is generally <u>Delta</u> connected, and the secondary sides can either be <u>Delta</u> or <u>Wye</u> connected.

4. The Delta-Delta (Δ-Δ) connection is used to deliver a single voltage on the secondary side. The connections to both sides will be with three conductor cables.

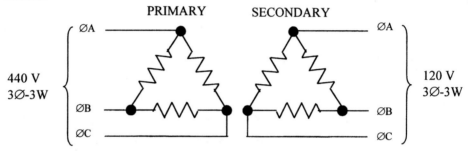

Diagram of a 440/120V-3∅ Δ-Δ

FIGURE B8-2

5. The Delta-Wye (Δ-Y) connection is used to deliver a <u>dual</u> voltage on the secondary side. The primary connection will be with a three conductor cable, and the secondary will be with a four conductor cable.

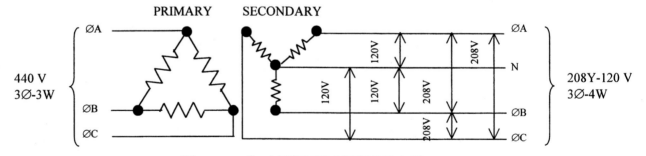

Diagram of a 440/208Y-120V-3∅ Δ-Y

FIGURE B8-3

6. Since we are using a Δ-Δ connected transformer bank, from step 4 above we know the secondary cables must be three conductor. We have already determined the secondary breaker to be a 400 amp trip with 3 poles.

7. Using the same Table A4-4 we can select a cable between 400 and 451 amps, which could be a TTNB-750 which carries 413 amps. This cable however has a very large diameter [1.385" per Table 8-15 of IEEE Std. 45-1998] making it very unwieldy to bend and handle in the confines of an engine room.

8. The minimum bending radius of all armored cables is six times the cable diameter [ABS Table 4/5B.2] requiring 8.31" of bend radius. To aid in installation, and ordering cable fittings, if a cable requirement is over 500 MCM, the circuit amps should be divided in half, and two cables selected and run in parallel.

9. Dividing the secondary amps trip of 400 by 2, we can select two cables which will each carry 200 amps. The secondary cables will be (2) TTNB-250's for each secondary transformer bank. Plug these figures into circuits P101(A) and P101(B).

IV.   EXAMPLE NO. 2:

Question 1: In Figure B8-4, if you had three (3) single phase 5 KVA transformers connected Delta-Wye (Δ-Y) to transform 440 Volts into 208Y/120 Volts – 3 phase, 4 wire, what would be the circuit breaker sizes for the primary and secondary sides?

Question 2: What cable sizes are required for the primary and secondary circuits?

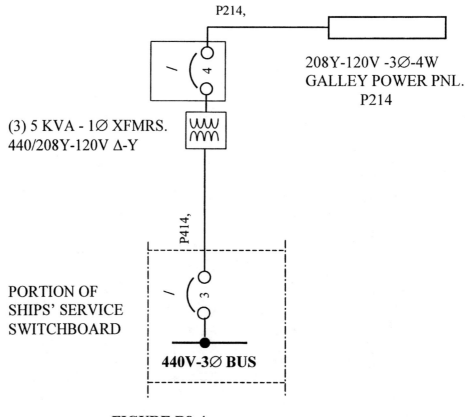

FIGURE B8-4

SOLUTION NO. 2:

A. First figure the total power of the transformer bank (3 x 5KVA = 15KVA).

B. Find the current of the primary side:

$$I = KVA \times 1000 \div 1.732 \times E$$
$$I = 15 \times 1000 \div 1.732 \times 440$$
$$I = 15000 \div 762.08$$
$$I = 19.68 \text{ amps primary current (round off to 20 amps)}.$$

C. Find the current of the secondary side. In dual voltage secondaries, the largest voltage is always used to calculate amps:

$$I = KVA \times 1000 \div 1.732 \times E$$
$$I = 15 \times 1000 \div 1.732 \times 208$$
$$I = 15000 \div 360.26$$
$$I = 41.63 \text{ amps secondary current (round off to 42 amps)}.$$

D. Size the circuit breakers based on 100% of primary and secondary currents, but not over 125%:

Primary is 20 x 1.25 = 25 amps
Secondary is 42 x 1.25 = 52.50 amps (round off to 53 amps)

E. The primary circuit breaker would be 100AF/20AT-3pole (Table A4-3). Plug this into the diagram.

F. The secondary circuit breaker would be 100AF/45AT. The circuit breaker for a Δ-Y configuration must be a 4 pole. Plug these into the diagram for the secondary circuit breaker.

G. Pick a primary three conductor cable which will carry between 20 and 25 amps. Table A4-4 shows a TTNB-14 carries 20 amps which is right on the borderline. To aid in voltage drop considerations later, this cable should be increased to the next higher cable of TTNB-12 which carries 24 amps. Since this is not over the 125% rating of 25, plug this cable into the diagram.

H. The secondary cable for a Delta-Wye configuration must be a four conductor. We can use the ampacity table for a three conductor given in Table A4-4 of Lesson 4. Due to the fourth conductor being a neutral, there is no need to derate this cable. We can use the "TTNB" chart and simply call it an "FTNB".

   1. A TTNB-7 is rated for 48 amps. This cable should be the one selected. Insert F̲TNB-7 into the diagram for the secondary cable.

TABLE B8-5 Standard Ships' Voltages [from IEEE Std. 45-1998 Articles 4.2 & 4.4]

EQUIPMENT	AC VOLTS			DC VOLTS
	SMALL SYSTEMS (TO 15KW)	INTERMEDIATE SYSTEMS (TO 100KW)	LARGE SYSTEMS (DUAL VOLT.)	
GENERATORS	120V	230V, 240V	450V, 480V. 600V, 690V	120V, 240V
DISTRIBUTION	115V	220V, 230V, 120V, 120/208V	440V, 460V, 575V, 660V, 120V, 230V, 120/208V	230V 115V
POWER	115V	220V, 230V	440V, 460, 575V, 660V	230V
LIGHTING	115V	120V, 120/208V	120V, 230V, 120/208V	115V

NOTE: For vessels having a very large electrical system requiring higher voltage power generation, consideration should be given to generating at 6600V, 4160V, or 2400V with some power utilization at 6000V, 4000V, or 2300V, three phase, respectively, with lower utilization voltages to be derived from transformers.

V. TRANSFORMER CIRCUITS – PROBLEM NO. 1

A. Size the circuit breakers and cables for the transformer circuits in Figure B8-6.
Main Lighting Transformer circuits will be P416.

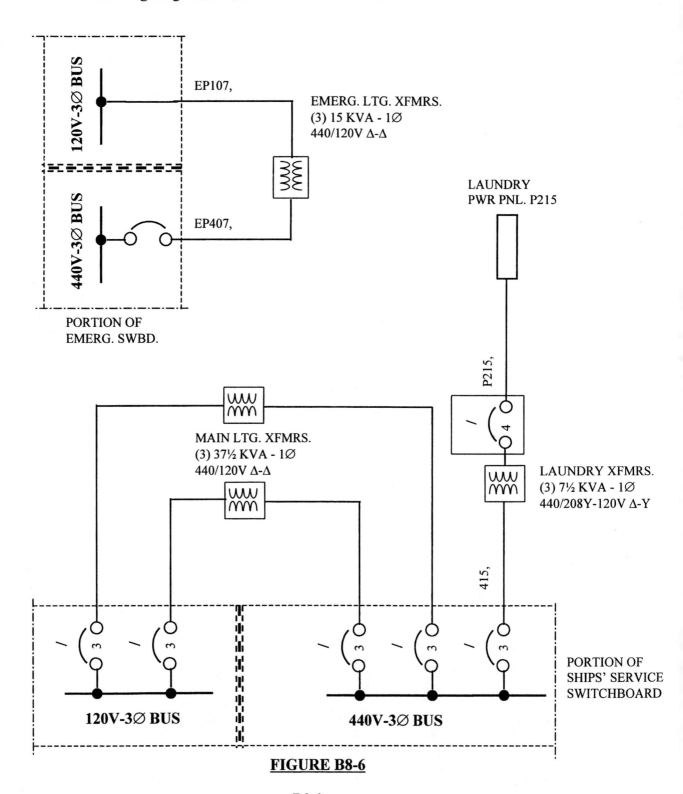

FIGURE B8-6

TABLE B8-7

Electrical Formulas for calculating Amperes, Kilowatts, KVA, and Horsepower

To Find	Direct Current	Alternating Current		
		Single Phase	2 phase*-Four Wire	Three Phase
Amperes when Horsepower is known	$\dfrac{HP \times 746}{E \times \% \, Eff.}$	$\dfrac{HP \times 746}{E \times \% \, Eff. \times P.F.}$	$\dfrac{HP \times 746}{2 \times E \times \% \, Eff. \times P.F.}$	$\dfrac{HP \times 746}{1.73 \times E \times \% \, Eff. \times P.F.}$
Amperes when Kilowatts Is known	$\dfrac{kw. \times 1000}{E}$	$\dfrac{kw \times 1000}{E \times P.F.}$	$\dfrac{kw. \times 1000}{2 \times E \times P.F.}$	$\dfrac{kw. \times 1000}{1.73 \times E \times P.F.}$
Amperes when Kv-a. Is known		$\dfrac{Kv\text{-}a. \times 1000}{E}$	$\dfrac{Kv\text{-}a. \times 1000}{2 \times E}$	$\dfrac{Kv\text{-}a. \times 1000}{1.73 \times E}$
Kilowatts	$\dfrac{I \times E}{1000}$	$\dfrac{I \times E \times P.F.}{1000}$	$\dfrac{I \times E \times 2 \times P.F.}{1000}$	$\dfrac{I \times E \times 1.73 \times P.F.}{1000}$
Kv-a.		$\dfrac{I \times E}{1000}$	$\dfrac{I \times E \times 2}{1000}$	$\dfrac{I \times E \times 1.73}{1000}$
Horsepower-(output)	$\dfrac{I \times E \times \% \, Eff.}{746}$	$\dfrac{I \times E \times \% \, Eff. \times P.F.}{746}$	$\dfrac{I \times E \times 2 \times \% \, Eff. \times P.F.}{746}$	$\dfrac{I \times E \times 1.73 \times \% \, Eff. \times P.F.}{746}$

I = Amperes; E = Volts; % Eff. = Percent Efficiency; P.F. = Power Factor;
kw = Kilowatts; Kv-a. = Kilovolt-amperes; HP = Horsepower

*For three-wire, two-phase circuits the current in the common conductor is 1.41 times that in either of the other two conductors.

The purpose of this lesson is to teach you the proper method of sizing generator and bus tie cables and circuit breakers.

I.   Figure B9-1 is a diagram of a typical ships power generating system showing a bus tie circuit between the main and emergency switchboards:

   A. All ocean vessels using electricity for ships' service power and lighting are provided with <u>at least</u> two ships service generating sets [USCG Subpart 111.10-3], with one set always acting as a standby, all connected to a ships service switchboard [USCG Subpart 111.10-4].

      1. Where the "at sea" load is greater than the capacity of a single generator, three or more sets will be arranged for paralleling together at the ships service switchboard. This allows more than one generator to give power to the switchboard in order to share the load, while always keeping one set as a standby.

      2. Thus if we had an "at sea" load of 900 Kilowatts, we could use three (3) generators rated 500 Kilowatts each, all capable of being paralleled at the switchboard. Two of the sets would cover the load and one would act as standby.

   B. Passenger vessels, miscellaneous self-propelled vessels, cargo and tank ships, mobile offshore drilling units (MODU's), oceanographic vessels, and barges with sleeping accommodations for more than six persons, are also required to have an emergency source of power [USCG Table 112.05-5(a)].

Lesson 9

      1. This is usually an emergency generator, which is capable of supplying power for vital services in the event of loss of main ships power.

      2. This generator is connected to an emergency switchboard and has the capacity to carry the total <u>connected load</u> of this switchboard, both of which are always located in the same space above the main or weather deck [USCG Subpart 112.05-5].

   C. The emergency switchboard <u>normally</u> receives it's power from the main switchboard and ships' service generators through cables which connect the two switchboards. This is called the BUS TIE circuit [USCG Subpart 112.05-3].

      1. Acting through an automatic transfer switch on the emergency switchboard, the emergency generator is automatically started upon loss of normal ships power [USCG Subpart 112.25-3(b)].

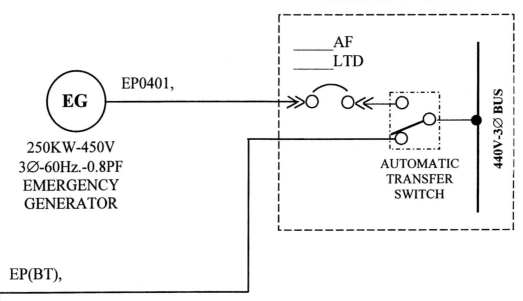

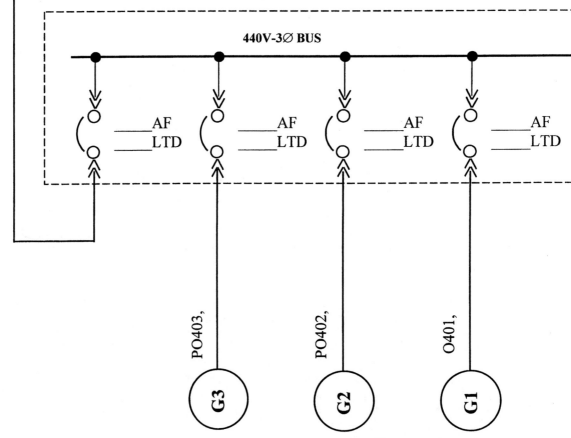

FIGURE B9-1

II.   U. S. Coast Guard rules for sizing circuit breakers and cables:

A. Each ships' service and emergency generator, and <u>any</u> generator arranged for paralleling, shall be protected by individual <u>trip-free air</u> circuit breakers [USCG Subpart 111.12-11(b)]. The pick-up setting of the long time overcurrent/delay trip (L.T.D.) shall not exceed 115% of the generator rating [USCG Subpart 111.12(d) (1) and (2)].

B. Generator cables shall be rated to carry <u>at least</u> 115% of the generator rating, including any overload rating [USCG Subpart 111.12-9(a) (1) and (2)].

C. Ships service to emergency switchboard BUS TIE cables shall be rated to carry <u>at least</u> 115% of the emergency generator rating [USCG Table 111.60-7].

III.   <u>EXAMPLE NO. 1:</u>

Question 1: As shown in Figure B9-1 we have a main switchboard supplied by three (3) generators, an emergency switchboard supplied by its generator, with both switchboards being connected together through a BUS TIE circuit. What are the cable and circuit breakers required?

<u>SOLUTION NO. 1:</u>

A. The first step is to determine the current rating of each generator using the formula for three phase:

$$I = KW \times 1000 \div E \times P.F. \times \sqrt{3}$$

1. Substituting values for the ships service generators:

   $I = 500 \times 1000 \div 450 \times 0.8 \times 1.732$
   $I = 500000 \div 623.52$
   $I = \underline{801.89}$ Amps Full Load

2. Substituting values for the emergency generator:

   $I = 250 \times 1000 \div 450 \times 0.8 \times 1.732$
   $I = 250000 \div 623.52$
   $I = \underline{400.94}$ Amps Full Load

B. The next step is to multiply the full load currents by 115%:

1. Ships service generators;
   801.89 x 1.15 = 922.17 (round off to <u>922</u> Amps).

2. Emergency generator:
   400.94 x 1.15 = 461.08 (round off to <u>461</u> Amps).

C. Now we can select the cables and circuit breakers for the generators. You will notice the difference in the circuit breaker symbols shown in Figure B9-1,

1. All the circuit breakers shown in previous lessons have been molded case <u>thermal</u> types with standard trip ratings that operate on the cable temperatures.

2. Since the rules state that our generators must have air circuit breakers with long time delays (L.T.D.), we show the symbol differently.

D. Ships Service Generators

1. We have already determined that the ships service generators are rated at 801.89 amps with an overload rating of 922 amps. A review of Table A4-3 of Lesson 4 reveals the Frame size will be 1800 AF, but there is no thermal breaker trip rating between 800 and 1000 amps.

2. In the case of generator breakers, however, the rules state this breaker must be an adjustable breaker capable of being set or adjusted to closely match the generator capabilities. This breaker should be set between the full load and overload rating. Show this trip rating as 850 represented as a long-time delay (L.T.D.). Plug this value into the circuit breaker symbol for all three (3) ships service generators (1800AF – 850 LTD – 3 pole).

3. Now select a cable that will carry <u>at least</u> the overload rating of 922 amps.

4. Since our load is above the ampacity of any single three conductor cable in Table A4-4 of Lesson 4, we must divide the load and select two equal cables capable of carrying half the load each (922 ÷2 = 461 amps).

5. Since this is still above any cables in the chart we must divide the load in thirds (922 ÷ 3 = 307.33 amps) and select three cables which will each carry one-third of the load.

6. We can now select three (3) TTNB-500 which each carry 329 amps. This will total 987 amps which is well above the minimum rating of 922.

E. Emergency Generator

1. We have already determined that the emergency generator is rated at 400.94 amps with an overload rating of 461 amps. Using Table A4-3 of Lesson 4 for guidance we can select a circuit breaker LTD of 450, which is <u>between</u> these two values. This breaker has a Frame (physical) size of 1600. Plug these

figures in at the emergency switchboard breaker (1600AF - 450 LTD - 3 pole).

2. Select the cable that will carry <u>at least</u> the overload rating of 461 amps. Since the load is above the ampacity of the largest three conductor cable, split the load (461 ÷ 2 = 230.5 amps) and select two (2)TTNB-300's. Plug this into the cable from the emergency generator.

F. Bus Tie Circuit

1. The rules state that bus tie cables are to be rated for 115% of the emergency generator circuit.

2. The bus tie cable is connected to the emergency switchboard, which is always located in the same space as the emergency generator. Therefore the bus tie cable will always be rated at a 45^0C engine room ambient temperature no matter where else the cable may run.

3. We can see by the foregoing that the bus tie cable will always be rated the same way as the emergency generator cable. All we have to do for the bus tie circuit is duplicate the emergency generator circuit.

4. Plug in the same breaker (1600AF – 450AT – 3 pole) and cable ((2)TTNB-300)for the bus tie as was used for the emergency generator.

IV.   <u>EXAMPLE NO. 2:</u>

Question 1: In Figure B9-2, what size cables and circuit breakers for the generators and bus tie circuit would be required?

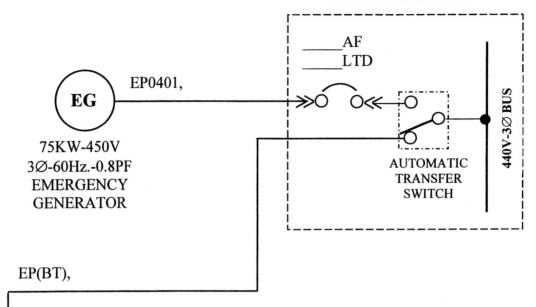

PORTION OF
EMERGENCY SWBD.

EP0401,

75KW-450V
3∅-60Hz.-0.8PF
EMERGENCY
GENERATOR

AUTOMATIC
TRANSFER
SWITCH

440V-3∅ BUS

EP(BT),

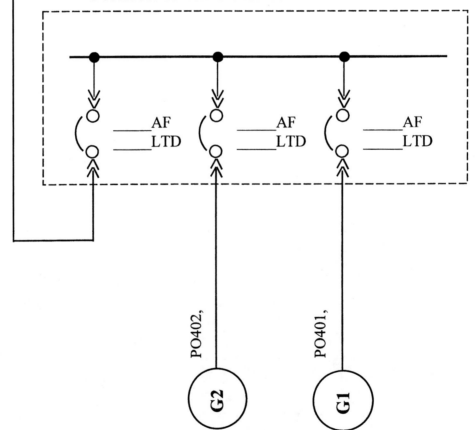

PORTION OF SHIPS SERVICE SWITCHBOARD

PO402,   PO401,

(2)300KW-450V-3∅-60Hz.-0.8PF
SHIPS' SERVICE GENERATORS

FIGURE B9-2

SOLUTION NO. 2:

A. Calculate the amps for each generator. Remember that the power factor for ships machinery is usually 0.8.

1. Ships service generators:

   I = 300 x 1000 ÷450 x 0.8 x 1.732
   I = 300000 ÷623.52
   I = <u>481.13</u> Amps x 115% = <u>553</u> Amps

2. Emergency generator:

   I = 75 x 1000 ÷450 x 0.8 x 1.732
   I = 75000 ÷623.52
   I = <u>120.89</u> Amps x 115% = <u>138</u> Amps

B. Select air operated adjustable circuit breakers with long time delays based on trip setting between rated and overload amps.

1. A 500 amp trip is between 481 and 533 amps, therefore the ships service generator breakers are 1600AF – 500AT – 3 pole.
2. A 125 amp trip is between 120 and 138 amps, therefore the emergency generator breaker is a 225AF = 125AT – 3 pole.

C. Select cables which carry at least the 115% amp rating of the generators, based on 45^{0}C ambient temperature.

1. Two (2) TTNB-400 cables each carry 286 amps which is under half the rated overload value (276.5 amps) for the ships service generators. Use (2)TTNB-400 which each carry 286 amps.
2. A TTNB-2/0 carries 145 amps at 45^{0}C which is above the overload rating of 138 amps. Use this cable for the emergency generator circuit.

D. The bus tie cable and circuit breaker will be the same size used for the emergency generator, since all rules are identical for the two circuits.

V.   GENERATOR AND BUS TIE CIRCUITS - PROBLEMS

A. We have a vessel with an electric power plant consisting of three (3) 900 KW – 450 Volt - 3∅ - 60 Hertz ships service generators, a 300 KW – 450 Volt - 3∅ - 60 Hertz emergency generator, and automatic switching between the main and emergency switchboard through a BUS TIE circuit.

1. What is the total generating capacity of the vessel? _____KW

2. What would be the largest allowable "at sea" load? _____KW

3. What size circuit breaker would be required for the :

(a) Ships' Service Generators   _____

(b) Emergency Generator   _____

(c) Bus Tie Circuit   _____

4. What size cables would be required for the:

(d) Ships' Service Generators   _____

(e) Emergency Generator   _____

(f) Bus Tie Circuit   _____

5. Make a One-Line Diagram depicting the above.

The purpose of this lesson is to teach the installer or designer how to properly select motor controllers.

I.   We have learned that a cable carries electrical power to a motor, and that a fuse or circuit breaker protects this cable from overheating.

    A. But what protects the motor itself from an overload or malfunction?

    B. The answer is the starter, or motor controller, which has sensing elements built into it for automatically disconnecting the motor from its' source of power.

II.   Motors which do not require controllers:

    A. Motors which are rated at 1.5KW (2 horsepower) or less, and 250 Volts or less, do not need to have motor controllers. A disconnect switch may be used instead, as long as the switch is rated at least twice the full load current of the motor [ABS Rule 4/5C4.17.2 and IEEE Std. 45-1998 Sub-clause 17.5].

    B. Motors which are rated at 1/8 horsepower or less, that are normally left running, such as clock motors and the like, don't even need the disconnect switch. The fuse or circuit breaker on the panel supplying them are sufficient protection, since these motors are not subject to overloads [NEC 1999 Section 430.81(b)].

    C. Portable motors rated 1/3 horsepower or less, such as refrigerators, mixers, slicers, and the like, may use plugs and receptacles as the disconnecting means [NEC 1999 Section 430.81(c).

Lesson 10

III.   Motors which require controllers:

    A. Any motor rated over 250 volts, and/or any motor larger than 1.5KW (2 horsepower) should have a controller capable of starting and stopping the motor [ABS Rule 4/5C4.17.2].

IV.   Selecting the proper controller:

    A. Motor controllers are <u>typed</u> depending on the use intended for the motor.

        1. Running protection for low voltage must be provided for any motor over 0.5 KW [ABS Rule 4/5A5.13.4]. For most motors this <u>type</u> of controller shall be Low Voltage Protection (LVP) [ABS Rule 4/5A5.13.5].

           (a) This means that if power is lost to the motor, it will shut down and will not start-up again by itself when power is restored. Low Voltage Protection <u>protects</u> the operator.

2. Low Voltage Release (LVR) types shall only be provided on auxiliaries which are vital to the operations of the ships propelling equipment, where automatic restarting would not be a hazard [ABS Rule 4/5A5.13.5]. This also includes fire pumps and elevator motors [USCG Subpart 111.70-3(b)]. Steering gear motors are always typed as Low Voltage Release [ABS Rule 4/5A6.3.2 and USCG Subpart 111.70-3(b)].

   a) This means that if power is lost to the motor it will shut down, but will automatically start up again when power is restored. Low Voltage Release <u>releases</u> the operator from the responsibility of restarting the motor manually when power is restored.

   b) Steering gear circuits, in addition to having Low Voltage Release controllers, cannot use circuit breakers with long time overcurrent protection. The circuit breakers must have <u>Instantaneous Trip</u> (IT) only, rated at least twice the full load current of the motor, and be arranged to permit the passage of the appropriate starting currents [ABS Rule 4/5A6.3]. This instantaneous trip must be set to trip at not less than 200% of the locked rotor current of one motor, plus any other loads that may be on the feeder [IEEE Std. 45-1998 Sub-clause 18.3.2]. Table B10-4 of this lesson gives motor locked rotor amps for motors up to 200 horsepower.

   c) Low Voltage Protection and Low Voltage Release controllers utilize <u>full voltage</u> (FV) for starting.

3. To prevent severe power draws on generators, large motors (above 200 HP) usually use controllers which have auto-transformers in them to furnish Reduced Voltage (RED.V.) for starting [IEEE Std. 45-1998 Sub-clause 34.3]. These Reduced voltage starters are much larger physically than full voltage starters.

B. A motor controller <u>size</u> is rated at not less than the horsepower rating of the motor.

1. This size is based on motor voltage, phases, and horsepower and is stated as a NEMA (National Electrical Manufacturers Association) size. See Table B10-2 at the end of this lesson for controller sizes.

C. A motor controller <u>enclosure</u> is based on its physical location on the vessel:

1. Enclosures for motor controllers or motor control centers must meet certain NEMA or IEC requirements [USCG Subpart 111.70-3(a)].

   a) Controllers in wet locations such as the open deck, refrigerated spaces, cargo spaces, etc., should have <u>watertight</u> enclosures [IEEE 45 Std. 45-1998 Sub-clause 17.4a].

b) Controllers mounted in main or auxiliary machinery spaces, or similar spaces adjacent to a weather deck access, where the equipment may be subject to splashing, mechanical damage, or dripping fluids should have splashproof or dripproof enclosures [IEEE 45 Std. 45-1998 Sub-clause 17.4b].

c) In all interior locations where controllers may be exposed to dripping fluids or condensation they should be dripproof.[IEEE Std. 45-1998 Sub-clause 17.4c].

d) In some cases controllers may be required in hazardous locations. Requirements for this equipment must comply with either the NEC (National Electrical Code) or the IEC (International Electrotechnical Commission) [USCG Subpart 111.105-3].

   While the NEC and the IEC requirements are similar in some respects, there are differences enough to preclude interchangeability of components in many instances. The acceptability of hazardous location electrical equipment is determined by the appropriate regulatory bodies and classification societies [IEEE Std. 45-1998 Annex A].

   These Hazardous Locations are listed by the NEC as follows:
   Class I: Flammable gases, Class II: Combustible dusts, and Class III: Ignitable fibers or flyings. These are further subdivided into four Groups and two Divisions[NEC 1999 Section 500-1].

   The IEC lists these differently as"
   Group I: Mining, and Group II: Surface industries (including offshore). These are further subdivided into three subgroups and three zones, plus two other zones for combustible dust.

   Depending on the classification of the space, the motor controllers required may be:

   > Explosion proof
   > Dust Ignitionproof
   > Dusttight

e) Explosion Proof controllers are those enclosed in a case which is capable of withstanding an explosion of a specified gas or vapor inside itself, and preventing the ignition of a specified gas or vapor surrounding it. These controllers are designed to operate at a cool enough external temperature that surrounding gases or vapors cannot be ignited. Due to their construction, explosion proof equipment cannot be made watertight, and should never be put in a space classed for watertight equipment.

D. Determine what control functions are required:

1. A controller may be located adjacent to the motor it serves, and is capable of being mechanically locked in the "open" position. If it is desired to group the controllers at a central location, or for some other reason it is not feasible to locate the controller within sight of (or fifty feet away from) its motor, a means of locking the controller in the open position must be provided [ABS Rule 4/5B2.13.2 and IEEE Std. 45-1998 Sub-clause 17.5].

    a) In some cases it is desired to have controllers grouped in a central location. This is accomplished by motor control centers (MCCs). These have modular units that are stacked and lined up side by side to present a neat and convenient means of controlling groups of motors from a central location, all being supplied by a single feeder. This feeder cable must be rated at 100% of all connected motor full load amps [ABS Rule 4/5A4.1.4].

    b) If motor control centers are used, each individual controller in the group is selected the same as if it was by itself. In this case a power panel would not be used to supply the motors.

    c) The motor control center itself will have individual circuit breakers for each motor; but instead of thermal type trips, the circuit breakers will have instantaneous trips (IT) [ABS Rule 4/5A5.13.1]. These must be set to trip at values not exceeding 1300% of the motor full load currents [IEEE Std. 45-1998 Sub-clause 17.1].

2. The motor controllers, besides protecting the motor from overloads, can also be ordered with characteristics to operate the motor from remotely located pushbuttons, pressure switches, temperature switches, level sensors, and to supply power to a heater within the motor casing.

    a) This function requires a selector switch for "Hand – Off – Automatic" operation. Control circuits are usually rated at 115 Volts, single phase. The controller can be ordered to include a control transformer which will convert the incoming voltage to the control voltage.

E. Some other controller functions:

1. Some motors, such as capstans will be rated at more than one speed, which requires a pushbutton or selector switch to select "Fast – Slow – Stop" modes. The controller would be ordered as a multi-speed type.

2. Some other motors, such as cargo winches, topping winches, vang winches, have the ability to be run in either direction. In this case a pushbutton or selector switch would be furnished having "Fwd – Rev – Stop" modes. The controller would be ordered as a reversing type.

V.   <u>EXAMPLE NO. 1:</u>

Question 1: As shown in Figure B10-1, we have a 100 horsepower Bilge Pump with a motor heater, which is automatically started and stopped by high and low level switches. Since there is no pushbutton at the motor we can assume that the controller is within sight of the motor. How would we select the proper controller?

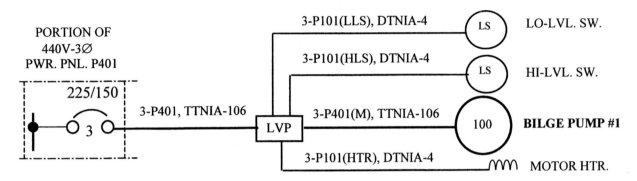

<u>FIGURE B10-1</u>

<u>SOLUTION NO. 1:</u>

A. The first step is to determine whether or not a controller is even necessary. Since the motor is above 2 horsepower and the voltage is above 250, we know we need a controller.

B. Using Motor Controller Worksheet B10-6, we can proceed step-by-step in selecting the proper controller as follows:

(1) Fill in the purpose of the motor (Bilge Pump No. 1).

(2) Fill in the Motor Horsepower (100), voltage (440), Phase (3), Cycles (60), Power Factor (0.8), Full Load Amps (124 – from Table A5-2 of Lesson 5), and Locked Rotor Amps (738 – from Table B10-4 of this Lesson).

(3) Put a check by the motor direction (Non-reversing). We know that a bilge pump will have a non-reversing motor. We do not want the motor to be run in both directions. A reversing motor is usually for winches and the like.

(4) Check which motor speed we have (single). If the motor is single speed it will have a single speed controller. Two speed and other speeds are for motors such as cargo, vang, and topping winches, capstans, and the like. These are usually line handling motors.

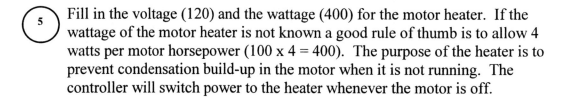

(5) Fill in the voltage (120) and the wattage (400) for the motor heater. If the wattage of the motor heater is not known a good rule of thumb is to allow 4 watts per motor horsepower (100 x 4 = 400). The purpose of the heater is to prevent condensation build-up in the motor when it is not running. The controller will switch power to the heater whenever the motor is off.

There should always be a motor heater installed in motors mounted in the weather or in wet classed spaces. The voltage to the heater can be either 440 or 120. If the voltage is 440 then the heater will usually get its power through the same cable that supplies the motor with power. If the voltage is 120 the heater will get its power through the control transformer in the controller, and a separate cable is required from the controller to the motor.

(6) Check which type of controller is required (Low Voltage Protection). We know that the Low Voltage Release is used only for steering gear and certain vital and auxiliary propulsion motors.

(7) Check which starting voltage will be used (Full Voltage). Since our motor is not large (over 200 HP) we do not need to use a Reduced Voltage Starter.

(8) List the controller size (NEMA 4) from manufacturers catalogs, or Table B10-2 of this Lesson.

(9) List the controller location (Engine room) and put a check by "Within Sight of Motor".

(10) Check the location rating (Splashproof).

(11) Fill in the enclosure type (NEMA 3S) from Table B10-3 of this Lesson.

(12) Check which control functions are required on the door of the controller. Since our diagram shows the motor started and stopped by level switches, we must have a "Hand – Off – Automatic" selector switch in the door.

This enables the pump to be operated manually when the switch is turned to "Hand". When it is desired to have the level switches start and stop the pump the switch is turned to "Auto".

Under other we will need to list a "lock on stop" for the selector switch. This enables the motor to be locked in the open position, to prevent its being energized when it is being serviced.

We do not need "Stop" or "Run" indicating lights in the controller door unless it is not located within sight of the motor, unless so requested by ships specifications.

(13) Put a check by the auxiliary contacts required for remote control (High Level and Low Level Switch).

(14) List the primary voltage (440) and the secondary voltage (115) for the control transformer. The primary voltage is always the same voltage as the motor. The secondary is usually 115 volts, which is common for motor control circuits.

TABLE B10-2 – Motor Controller Sizes

Motor HP	NEMA Size	Motor HP	NEMA Size
0 to 2	00	50.1 to 100	4
2.1 to 5	0	100.1 to 200	5
5.1 to 10	1	200.1 to 400	6
10.1 to 25	2	400.1 to 600	7
25.1 to 50	3	600.1 to 900	8

TABLE B10-3 – Common NEMA Enclosures [NEMA 250-1991 Section 2]

NEMA Type	IEC Design.	Description
NEMA 1	IP11	General Purpose / Semi-guarded
NEMA 1	IP21	Guarded
NEMA 2 / 12	IP32	Dripproof
NEMA 3S	IP54	Splashproof
NEMA 4 / 4X	IP66	Dusttight, Waterproof
NEMA 5	IP54	Dustproof
NEMA 6	IP67	Watertight
NEMA 6P	IP68	Submersible
NEMA 13	IP52	Oilproof, Oiltight

TABLE B10-4 – Locked Rotor Current For Motors

Max. HP Rating	Motor Locked-Rotor Current Amperes						
	Single Phase		Two or Three Phase				
	115V	230V	115V	200V	230V	460V	575V
½	58.8	29.4	24	14	12	6	4.8
¾	82.8	41.4	33.6	19	16.8	8.4	6.6
1	96	48	42	24	21	10.8	8.4
1 ½	120	60	60	34	30	15	12
2	144	72	78	45	39	19.8	15.6
3	204	102	-----	62	54	27	24
5	336	168	-----	103	90	45	36
7 ½	480	240	-----	152	132	66	54
10	600	300	-----	186	162	84	66
15	-----	-----	-----	276	240	120	96
20	-----	-----	-----	359	312	156	126
25	-----	-----	-----	442	384	192	157
30	-----	-----	-----	538	468	234	186
40	-----	-----	-----	718	624	312	246
50	-----	-----	-----	862	750	378	300
60	-----	-----	-----	1035	900	450	360
75	-----	-----	-----	1276	1110	558	444
100	-----	-----	-----	1697	1476	738	588
125	-----	-----	-----	2139	1860	930	744
150	-----	-----	-----	2484	2160	1080	864
200	-----	-----	-----	3312	2880	1440	1152

VI.   MOTOR CONTROLLERS – PROBLEMS

A. Using blank Motor Controller Worksheets B10-6, select the three motor controllers in Figure B10-5:

1. The Steering Gear motor with adjacent controller has controls remotely mounted in a control console.

2. The Air Compressor with adjacent controller is automatically controlled by a dual pressure switch.

3. The Capstan is mounted in the weather with an adjacent push button. The passageway on the deck below holds the motor controller.

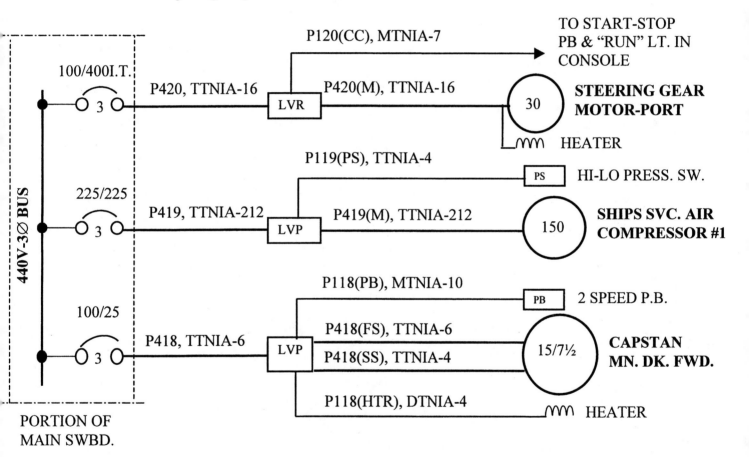

FIGURE B10-5

MOTOR CONTROLLER WORKSHEET **B10-6**

(1)

MOTOR DATA

Motor use _____

(2)

Motor horsepower _____ Voltage _____ Phases _____ Cycles _____

(3)

   Power Factor _____ Full Load Amps _____ Locked Rotor Amps _____

Motor direction: Reversing _____   Non-Reversing _____

(4)

Motor speed: Single _____ Two Speed _____ Other _____

(5)

Motor heater: Voltage _____ Wattage _____

CONTROLLER DATA

(6)

Controller Type: Low Voltage Protection (LVP) _____
                 Low Voltage Release (LVR) _____

(7)

Starting Type: Full voltage (FV) _____
            Reduced Voltage (RED.V.) _____

(8)

Controller Size: NEMA _____

(9)

Controller Location _____
In sight (or within 50') of Motor _____ Not in sight of Motor _____

(10)

Location Rating: Watertight _____ Splashproof _____
                Dripproof _____ Explosion Proof _____

(11)

Enclosure Type: NEMA _____

(12)

Door Mounted Controls: Start-Stop Pushbutton with Lock on Stop _____
   Start-Stop Pushbutton _____   "Hand-Off-Automatic" Selector Switch _____
   "Fwd-Rev-Stop" Pushbutton _____   "Fast-Slow-Stop" Pushbutton _____
   Motor "Run" Light _____   Motor "Stop" Light _____
   Other _____   None _____

(13)

Auxiliary Contacts for Remote: Start-Stop Pushbutton with Lock on Stop _____
   Start-Stop Pushbutton _____   "Hand-Off-Automatic" Selector Switch _____
   "Fwd-Rev-Stop" Pushbutton _____   "Fast-Slow-Stop" Pushbutton _____
   Motor "Run" Light _____   Motor "Stop" Light _____
   High Level Switch _____   Low Level Switch _____
   High Temp. Switch _____   Low Temp. Switch _____
   High Press. Switch _____   Low Press. Switch _____
   Other _____   None _____

(14)

Control Transformer: Primary Voltage _____ Secondary Voltage _____

MARINE ELECTRICAL BASICS

WORKBOOK

Module C

ADVANCED

I. When someone wishes to have a ship built, a Naval Architect is usually contacted to draw up a set of "Contract Plans and Specifications", which reflects the owners intentions.

This set of specifications is many times based on the Maritime Administration's (MARAD) "Standard Specification for Ship Construction". Although the particulars may vary greatly from vessel to vessel, and many items may be added or omitted from the guidance standard, the means of electrical distribution generally follows a set pattern. These specifications usually state the design and installation requirements to be "in accordance with ABS, USCG, and IEEE-45 regulations", or whatever local or country rules may be applied.

The Marine Electrical Designer should be able to take the Contract Plans and Specifications and draw up a One-Line Diagram.

II. The One-Line Diagram is the <u>key</u> electrical plan for a vessel, and should be designed to give as much information as possible. The name (One-Line) is derived from representing cables and bus bars with one line on the drawing, no matter how many conductors the cables have, or how many bus bars are in a switchboard or panel.

 A. The One-Line Diagram is <u>required</u> [USCG Subpart 110.25-1(a) and ABS Rule 4/5A1.1.2] to show at least the following:

  1. The type and size of generators and prime movers.
  2. The type and size of generator cables, bus-tie cables, feeders, and branch circuit cables.
  3. Power, lighting, and Interior Communication panelboards with the number of circuits, and the rating of each energy consuming device.
  4. The type and capacity of storage batteries.
  5. The rating and interrupting capacity of circuit breakers, and the rating or setting of switches and overcurrent devices.

Lesson II

III. The sequence of power distribution:

 A. Figure C11-1, in very simple form, shows a typical distribution system as represented on a One-Line Diagram.

 B. By following the diagram it can be seen that power is distributed from:

  ① The ships Service Generators to the Main Switchboard power section.

  ② From Shore Power to the main Switchboard power section.

  ③ From the Main Switchboard power section to the main switchboard lighting section via a step-down transformer.

4 From the Emergency Generator to the Emergency Switchboard power section.

5 From the emergency Switchboard power section to the emergency switchboard lighting section via a step-down transformer.

6 From the Main Switchboard power section to the Emergency Switchboard power section through a bus-tie circuit and an Automatic Transfer Switch.

7 From the distribution sections of either switchboard:

7a To a branch circuit for an individual controller and motor.

7b To a Power Panel.

7c From a Power Panel, to branch circuits for individual controllers and motors.

7d To a lighting branch circuit.

7e To a Lighting Panel, or to an Interior Communication Panel, then to their lighting or I.C. branch circuits.

7f To a Power or Lighting Distribution Panel, serving to subdivide the feeder to sub-feeders supplying other Power or Lighting Panels and their branch circuits.

7g To a transformer, then to a Power or Lighting Distribution Panel, serving to subdivide the feeder to sub-feeders supplying other Power or Lighting Panels and their branch circuits.

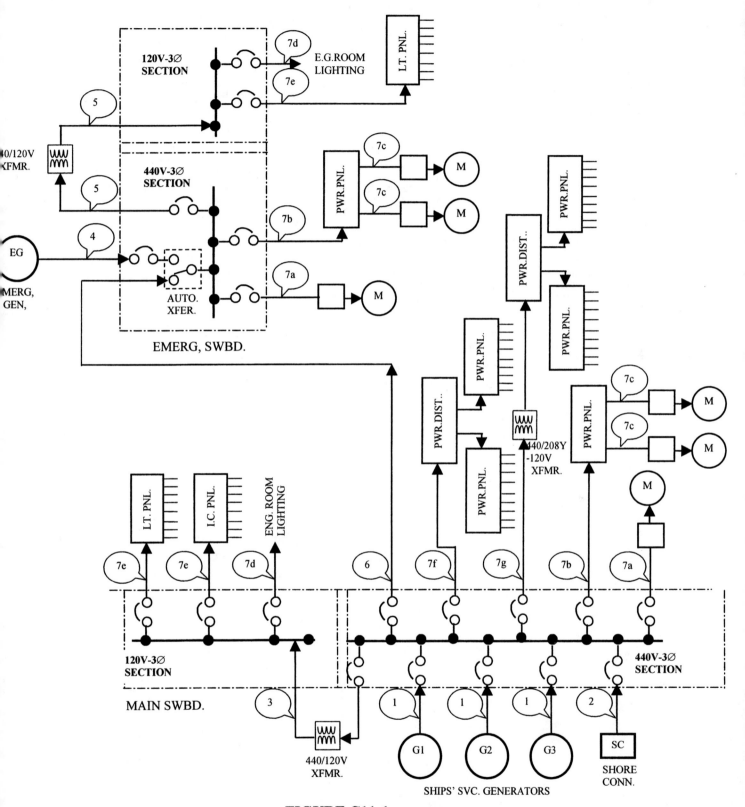

FIGURE C11-1

IV.   Circuit Designations:

    A. Panelboards have to be provided with a directory giving the <u>circuit designation</u>, load, and trip setting for each circuit [USCG Subpart 111.40-11].

    B. Switchboards and motor control centers shall have nameplates provided for each piece of apparatus to indicate clearly its purpose [USCG 111.30-15]. Nameplates for motor control center branch circuits should include the <u>circuit designation</u> [IEEE Std. 45-1998 Sub-clause 17.2].

    C. All electrical circuits, including power, lighting, communications, control, and electronics systems, should be defined by <u>appropriate designations</u> [IEEE Std. 45-1998 Annex B]. Some of the more commonly used designations are listed in Table C11-9 of this Lesson.

    D. From the above it can be seen that it is necessary to set up a system of identifying the various power, lighting, and communications system cables. If the system of identification is set up properly it will also aid the installing electrician in identifying the purpose of a particular cable simply by knowing its designation.

    E. Power and Lighting cable designations are made up of letters and numbers signifying the service, and cable type and size.

        1. Punctuation is very important on a One-Line Diagram and should never be neglected. As shown in Figure C11-2:

            • the letters and numbers to the left of the comma are the cable designation, which reveals the service, voltage, and circuit number.

            • The letters and numbers to the right of the comma are the cable type and size, which reveals the number of conductors, the insulation type, the jacket, the armor, and the circular mil size.

            • When selecting a cable, the manufacturers charts should always be consulted to pick the proper type and size.

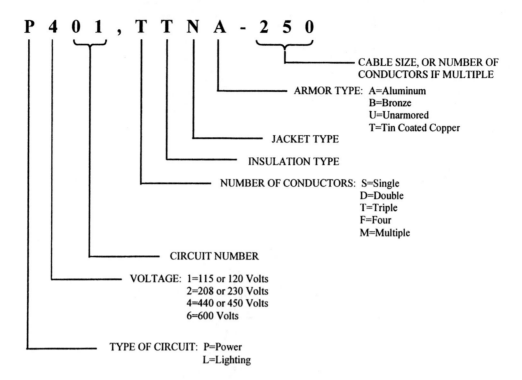

FIGURE C11-2

2. Control cables for power and lighting can further be identified by adding in parenthesis the equipment to which they go, as shown in Figures C11-3, C11-4, and C11-5:

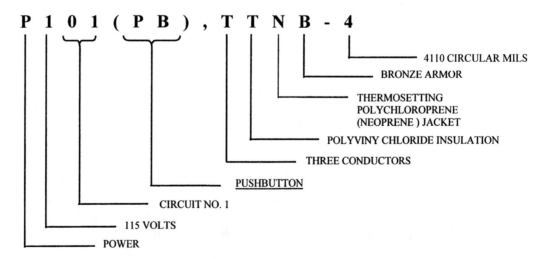

FIGURE C11-3

FIGURE C11-4

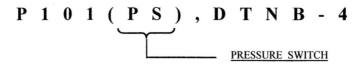

FIGURE C11-5

F. Control, signal, and communications cable designations are made up of letters and numbers signifying the system to which they belong to the left of the comma, and the cable type and size to the right of the comma.

1. An example of this would be the cable between two sound powered telephones as shown in Figure C11-6:

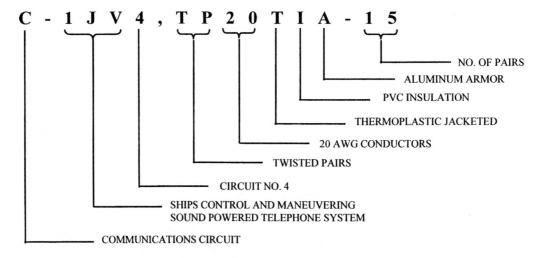

FIGURE C11-6

2. Most communications and signal system circuit designations are preceded by a letter. The most commonly used ones are "C" for Interior Communication, "K" for Control, or "R" for Electronics. Figures C11-7 and C11-8 show some examples of this.

FIGURE C11-7

FIGURE C11-8

G. Circuit designations are also important in order to label any connection box concealed behind paneling or sheathing since these should be identified on the vessel [ABS Rule 4/5B3.33]. Table C11-9 of this lesson lists some of the more commonly used circuit designations for signal and communications systems.

V. When setting up a One-Line Diagram, the following basic rules should be kept in mind:

A. The U.S. Coast Guard and the American Bureau of Shipping are very specific about what loads must be supplied from the emergency switchboard. When setting up a distribution system these regulations should be consulted for what rules apply to your emergency source of supply.

   1. In general, the following is always connected to the emergency switchboard if installed or required on the vessel [USCG Subpart 112.15-1, 112.15-5, and ABS Rule 4/5A3.3]:

   a) Navigation lights
   b) A sufficient number of lights in machinery spaces for emergency work.
   c) Sufficient lighting throughout accommodations, public spaces, and machinery rooms to allow passengers and crew to find their way out.
   d) Exit signs.
   e) Illumination for safe operation of watertight doors.
   f) At least one light in each space where a person may be found such as public spaces, work spaces, machinery spaces, galleys, pump rooms, bow thruster rooms, paint and other stores lockers, steering gear rooms, windlass rooms, etc.
   g) Illumination for survival craft, including launching areas, muster stations, embarkation areas, and the water.
   h) Necessary electronic communications systems.
   i) Watertight door systems.

j) Essential communications systems.
k) Fire door holding and release systems.
l) Daylight signaling light.
m) Electric ships whistle.
n) Smoke detection system.
o) Fire and gas detection systems if installed.
p) Helicopter operations and deck area lighting, if installed.
q) Emergency systems required by SOLAS 74 such as:
- VHF Radio installation
- MF Radio installation
- Ship earth station
- MF/HF Radio station

r) Passenger elevators, if installed.
s) Emergency generator starting batteries.
t) General alarm system batteries.
u) One of the bilge pumps.
v) One of the fire pumps.
w) External lube oil pumps for propulsion and ships service generators, if necessary.
x) Each rudder angle indicator.
y) Each radio or Global Maritime Distress and Safety System (GMDSS).
z) Each radio direction finder, loran, radar, gyrocompass, depth sounder, Global Positioning System (GPS), speed log, rate-of-turn indicator, and propeller pitch indicator.

2. Other loads may be added if authorized, but only if the emergency source of power is large enough to carry 100% of the connected load [USCG Subpart 112.05-1(c)(1)]. All loads are normally powered from the main switchboard through an automatic transfer switch on the emergency switchboard [USCG Subpart 112.25-3].

B. When there are two independent steering systems, they need two separate feeders direct from a switchboard (s). One of the feeders may be from the emergency switchboard if the source of emergency power is large enough [USCG Subpart 112.15-5(k) and ABS Rule 4/5A6.1].

C. In general, separate feeders from switchboards should be run for such groups as main and auxiliary machinery space loads, cargo handling gear, steering gear, navigation and radio systems, searchlights, and heating, ventilation, and air conditioning systems [IEEE Std. 45-1998 Sub-clause 11.18].

D. Separate feeders shall be run for cargo vent fans, machinery space vent fans, and accommodation spaces vent fans [ABS Rule 4/5A4.1.7 and IEEE Std. 45-1998 Sub-clause 11.18].

E. All vent fans must have a minimum of <u>two</u> means of stopping the fans in an emergency. One means can be the circuit breaker on the switchboard, but the other must be a remote pushbutton or the like, located outside of the space ventilated for machinery space fans, and located in the wheelhouse for cargo and accommodation space fans [USCG Subparts 111.103-1, 111.103-3 and ABS Rule 4/5A10.1.1].

F. Forced and induced draft fans, fuel oil transfer units and service pumps, and other similar fuel pumps need a remote stop pushbutton or the like outside of the space in which they are located [USCG Subpart 111.103-9 and ABS Rule 4/9.61.2].

G. Galley equipment has to have a means of disconnecting the power source in the same room as the equipment. If the power supply panel supplying the equipment is not in the same room, and the equipment is not served by a receptacle, a disconnect switch must be added adjacent to the equipment [IEEE Std. 45-1998 Sub-clause 18.6].

H. Navigation light indicator panels shall be protected by an overcurrent device (fused switch or circuit breaker) rated at twice the value of the panel main fuses. This device must be a circuit breaker on the emergency switchboard and may also be a fused switch located on the indicator panel. [USCG Subpart 111.75-17(a) and ABS Rule 4/5A7.3.3]. The panel must be supplied directly from the emergency switchboard, or by an unswitched through feed from the feeder supplying the wheelhouse emergency lighting panel [USCG Subpart 112.43-13].

I. Separate feeders shall be supplied for cargo space lighting. The distribution panels shall be outside the cargo spaces [ABS Rule 4/5A4.1.9 and IEEE Std. 45-1998 Sub-clause 11.16].

J. If lifeboat and liferaft floodlights are adjacent to each other, they shall be fed by separate circuits from the emergency switchboard [USCG Subpart 111.75-16(b)].

K. Fixed appliance branch circuits cannot have any lighting fixtures attached which are not an integral part of the appliance. A lighting panel will have nothing on it but lighting fixtures and receptacles, except desk or bracket fans, small heaters, motors of ¼ HP or less, and other portable appliances not rated over 190 watts [IEEE Std. 45-1998 Sub-clauses 11.23 and 11.25].

TABLE C11-9
Common Signal And Communications System Designations
[from IEEE Std. 45-1998 Annex B]

SYSTEM	DESIGNATION	SYSTEM	DESIGNATION
Announcing-Docking	C-8MC	Radio Receiving Ant. Dist.	R-RB
Announcing-General	C-1MC	Radio Receiving Ent. Dist.	R-RE
Announcing-Loudhailer	C-6MC	Refrig. Alarm Cargo	C-RH
Aux. M'ch'ry. Rm. Cont.	K-AC	Refrig. Alarm Ship Stores	C-RA
Call Bells	C-A	Rudder Angle Indicator	C-N
Central Alm. & Monitoring	C-AM	Salinity Indicator	C-SB
CO_2 Release Alarm	C-CO	Sewage Tk. Hi-Lvl. Alarm	C-STD
Cooling Water Hi-Temp. Alarm	C-EW	Shaft Revolution Indication	C-K
Echo Depth Sounder	R-SS	Smoke Indicator	C-SM
Electric Plant Control and Monitoring	K-EC	Steering Control	C-L
Emergency Gen. Set Control & Indication	K-EG	Steering Gear Alarm	C-LA
Engine Order Telegraph	C-MB	Tank Level Alarm	C-TD
Fire Detection and Alarm	C-F	Tank Level Indicator	C-TK
Fire Door Release	C-FR	Teleph.-Automatic Dial	C-J
Flooding Alarm	C-FD	Teleph.-Sound Pwr.-Call	C-E
Fog Alarm	C-FB	Teleph.-Sound Pwr.-Cargo & Ballast Control	C-6JV
Fuel Oil filling alarm	C-3TD	Teleph.-Sound Pwr.-Damage Control	C-JZ
F.O. Tank High-Level Alarm	C-4TD	Teleph.-Sound Pwr.-Engineers (Electrical)	C-5JV
General Alarm	C-G	Teleph.-Sound Pwr.-Engineers (Fueling)	C-3JV
Gyro Compass	C-LC	Teleph.-Sound Pwr.-Engineers (M'ch'ry Cont.)	C-2JV
Helm Angle Indicator	C-LH	Teleph.-Sound Pwr.-Misc.	C-7JV
Hospital and Nursing Call	C-AN	Teleph.-Sound Pwr.-Ship Cont. & Maneuvering	C-1JV
Lube Oil Lo-Level Alarm	C-EL	Underwater Log	C-Y
Lube Oil Lo-Press. Alarm	C-EC	Watertight Door Control	C-WD
Radar, Navigation	R-RN	Wet & Dry Bulb Temperature Indication	C-T
Radio Antenna	R-RA	Whistle Operator	C-W
Radio Direction Finder	R-RD	Wind Direction Indicator	C-HD
Radio GMDSS	R-GA	Wind Intensity Indicator	C-HE
Radio Receiver	R-RR		

I.   Most ships specifications list the type and amount of lighting required for various
        areas of a vessel. The intent is to ensure that adequate lighting is provided
        throughout a ship to suit the owner's method of operation.

        The "footcandles" required for each space may be listed as a part of the
        specifications, but if they are not; recommended levels of illumination have been
        established by the Illuminating Engineering Society of North America (IESNA), and
        are included as a part of this lesson.

II.   The U. S. Coast Guard only requires that each space used by passengers or crew
        must be fitted with lighting that provides for a safe habitable and working
        environment under normal conditions [USCG Subpart 111.75-1(c)].

        The Institute of Electrical and Electronics Engineers states that the illumination
        levels should be that recommended by the Illuminating Engineering Society (IES)
        [IEEE Std. 45-1998 Sub-clause 23.8.1].

III.   The purpose of this lesson is to give a step-by-step method for picking the quantity
        and type of fixture needed for each space on a ship based on the Zonal Cavity
        method. It is recommended that the IESNA "Recommended Practice for Marine
        Lighting" become familiar to the marine designer or installer.

IV.   First it is necessary for the designer or installer to have on hand the:

        • Ships' Specifications (to determine illumination levels required)
        • General Arrangement Plans (to determine physical square footages)
        • Inboard Profile (to determine room heights)

        A. Using the "ILLUMINATION CALCULATION WORKSHEET" C12-12
            provided with this lesson, and proceeding through the twelve steps which follow,
            it is possible to arrive at the type and quantity of fixtures required. A separate
            worksheet should be made for every different space on the vessel.

        B. Two examples of a completed worksheet are included as the final pages of this
            lesson to show two typical spaces on a ship: a paint locker requiring 15 foot
            candles of illumination, and a control room requiring 50 foot candles of
            illumination.

        (1) Fill in the Space Designation, Deck, Frames and Side of the ship from the
            General Arrangement Plans. There will be a separate worksheet for each
            different space. This will include passageways, store rooms, lockers, and all
            other illuminated spaces. If two spaces are similar, such as in many
            staterooms, only one worksheet need be made for the space.

Lesson 12

(2) Fill in the physical sizes of each space in feet and decimal equivalents of a Foot (Table C12-1). Deck heights may be taken from the Inboard Profile. Ceiling heights will be figured from the Joiner drawings (furniture arrangements and wall and ceiling finishes) or the joiner section of the Ships Specifications.

If there is no ceiling in the space, deck and ceiling heights will be the same. Odd shaped rooms will use average size dimensions.

TABLE C12-1 Decimal Equivalents Of A Foot

Inches	Feet	Inches	Feet	Inches	Feet
1	.08	5	.42	9	.75
2	.17	6	.50	10	.83
3	.25	7	.58	11	.92
4	.33	8	.67	12	1.00

(3) Determine the area (A) of the space in square feet by: A = Length x Width.

(4) (a) Determine the color of the finished surfaces of each space from the paint section of the ships specifications.

(b) From Table C12-2 list the percent of reflectance for each surface. This should be expressed in %, not in decimal form.

TABLE C12-2 Illumination Reflectance Percents For Surfaces

Surface Color	Reflectance Percent
White, very light to light tints of blue-green, cream, blue, buff or gray --	75%
Medium blue-green, yellow, medium buff, or gray -------------------	50%
Dark gray, medium blue --	30%
Dark blue, brown, dark gray, and many wood finishes, such as dark oak and mahogany ---	10%

(5) List the type of fixture selected. This should include the Wattage, Number of lamps, Fluorescent or Incandescent (type is usually called for in ships specifications). Manufacturers Name, Part Number, and Physical Size should also be listed if known. Reflectors should be selected based on the fixture height above the deck, per Table C12-3.

TABLE C12-3 Fixture Reflector Types

Height Above Deck	Reflector Type
12 feet or less ------------------------	Shallow Bowl Reflector
13 to 15 feet --------------------------	Shallow Dome Reflector
16 to 19 feet --------------------------	Deep Bowl Reflector
20 feet or more -----------------------	High Bay (Wide, Medium, Narrow)

6 List the lumen output of the fixture. As the lumen output varies from fixture to fixture, this should preferably be taken from the manufacturers data supplied with the fixture selected. If the fixture data is not available Table C12-4 may be used for estimating purposes. Final calculations should use the actual fixture data.

TABLE C12-4 Fixture Lumen Outputs

Lamp	Watts	Lumens	Life (Hours)
Incandescent	100	1670	750
(Inside frosted)	150	2760	750
	200 A-23	3940	750
	200 A-25	3820	750
	200 PS-25, PS-30	3680	750
	300 PS-25	6300	750
	300 PS-30	6000	750
	300 PS-35, PS-40	5850	750
	500	10500	1000
Fluorescent	40 (Rapid start)	3150	12000
(Cool white)	40 (Slimline)	2900	9000
	60 (800 ma.)	4050	9000
	90 (Bipin)	6300	9000
	110 (1500 ma.)	6900	7500
Quartz	500	10500	2000
Mercury	75	2600	16000
(Color improved)	100	4000	24000
	175	7800	24000
	250	11750	24000
	400	20500	24000
	1000	54000	24000
Metallic Additive	400	30000	7500
	1000	90000	7500
Ceramic Discharge	400	42000	7500

⑦ List the maintenance factor from ships specs. This figure will be the same for all spaces. The maintenance factor is a rating of how well maintenance is done and the conditions of operation. If not listed in the specifications use Table C12-5.

TABLE C12-5 Illumination Maintenance Factors

TYPE OF OPERATING CONDITIONS

Good – Clean air, free of fumes and dust, fixtures scheduled to be cleaned frequently, and lamps replaced systematically.

Medium – Less favorable atmospheric conditions, fixtures cleaned at fairly frequent intervals, and lamps replaced only after burnout.

Poor – Quite dirty locations and work atmospheres, spasmodic or poor maintenance of lighting equipment.

Lamp and fixture	Good	Medium	Poor
Incandescent	.75	.70	.65
Fluorescent	.85	.80	.75
Quartz	.75	.70	.65
Mercury	.65	.60	.55
Metallic Additive	.75	.70	.65
Ceramic Discharge	.75	.70	.65

⑧ List the foot candles required for each space as listed in the ships specs. If not listed, or if the values are less than that recommended by the I.E.S., use Table C12-6 on the following page.

TABLE C12-6 Illuminance Levels for Marine Lighting Tasks (From IESNA RP12-97 Table 3)

Area	Footcandles[a]	Area	Footcandles[a]
Living Areas		Ship's Offices [9]	
Cabins and Staterooms		General Lighting/Lobby Areas --------	20
General lighting ------------------------	10	Clerical Tasks -------------------------	75
Reading and Writing [4] [5] -------------	30	Wheelhouse, Pilothouse, Bridge	
Prolonged Seeing -----------------------	75	Day, In-port During Maintenance ------	20
Baths/Showers (general ltg.) ----------	20	Night when Underway [10]	0
Mirrors (personal grooming) -----------	50	Service areas	
Barber and Beauty shop		Food preparation	
General Lighting -----------------------	50	General -----------------------------	20
On Subject ----------------------------	100	Butcher shop ------------------------	75
Day Rooms		Galley -------------------------------	75
General Lighting [5] ---------------------	20	Pantry-------------------------------	30
Desks [4] [5] ------------------------------	50	Thaw room --------------------------	30
Dining Room, Messroom, Wardroom---		Scullery ------------------------------	30
Fine Dining	10	Food storage	
Coffee Shop, Cafeteria	20	Nonrefrigerated ---------------------	20
Snack, Fast Food	50	Refrigerated (ship's stores) ----------	10
Enclosed promenades		Laundries	
Night	8	General -----------------------------	30
Day	15	Machine, Pressing, Finishing, and	
Entrances and Passage Ways		Sorting (6)-------------------------	75
Night -----------------------------------	8	Sewing	150
Day -------------------------------------	15	Lockers -----------------------------	15
Daytime embarkation -------------------	30	Offices	
Gymnasiums		General -----------------------------	20
Exercise, Recreation --------------------	30	Reading -----------------------------	75
Hospital, Sick Bay		Passenger/Service Counters ------------	50
Dispensary (general lighting) ----------	30	Storerooms --------------------------	5
Doctor's Office -------------------------	30	Telephone exchange ------------------	75
Operating room		Operating Areas	
General lighting -----------------------	150	Access and Casing -------------------	15
Operating table -----------------------	2,000	Battery Room ------------------------	20
Wards		Boiler Rooms [6] ----------------------	20
General lighting -----------------------	10	Cargo Handling (weather decks) [6] ------	15
Critical Examination --------------------	100	Control stations (except navigating	
Reading --------------------------------	30	areas)	
Toilets ---------------------------------	20	General -----------------------------	20
Recreation areas		Control Consoles -----------------------	50
Ball rooms -----------------------------	15	Gauge and Control Boards -------------	30
Lounger, Cocktail lounges --------------	15	Switchboards -------------------------	30
Swimming pools		Engine Room [6] ----------------------	30
General Lighting -----------------------	15	Generator and Switchboard Room [6] ---	30
Underwater		Fan Room (ventilation & air	
Luminous Flux per Unit Surface		conditioning -------------------------	15
Area		Motor Room --------------------------	30
Indoor: 1,000 LM/M^2 (100 LM/FT^2)		Motor Generator Room (cargo	
Outdoor: 600 LM/M^2 (60 LM/FT^2)		handling) ----------------------------	15
Theater/Auditorium		Pump Room --------------------------	30
General -------------------------------	10	Shaft Alley --------------------------	10
During Program [5] ---------------------	0.1	Shaft Alley Escape Trunk --------------	3
Navigating Areas		Steering Gear Room [6] ---------------	30
Chart room		Windlass Rooms ---------------------	10
General ---------------------------------	15	Workshops	
On Chart Table -------------------------	75	General -----------------------------	30
Gyro room ------------------------------	20	On Bench Top ------------------------	75
Radar Room -----------------------------	20	Machine Shop [6] ---------------------	300
Radio Room		Cargo Holds	
Operating Areas ------------------------	75	Safety -------------------------------	3
Passenger Foyer -------------------------	30	During Cargo Handling ---------------	30
		Passageways and Trunks -------------	8

TABLE C12-6 Illuminance Levels for Marine Lighting Tasks (Continued)

NOTES:
  4. This task may be subject to veiling reflections.
  5. Illuminances may vary over a wide range depending upon the desired aesthetic effect in the space, the decorative scheme, and the specific activities taking place. Flexibility should be designed into the lighting system with multiple sources and controls.
  6. Supplementary lighting is recommended to provide design task illuminances.
  9. The recommended illuminances are for typical tasks and do not include special situations which may occur in practice. For detailed guidance please refer to the IESNA Lighting Handbook.
  10. At night the wheelhouse must be kept as dark as possible to permit visibility of objects on the water, hazards to navigation, and navigational lights.

(9)   (a) Determine the Ceiling Cavity Height (^{h}CC), the Room Cavity height (^{h}RC), and the Floor Cavity height (^{h}FC) based on Figure C12-7. Floor Cavity height is usually 2.5 feet. If the fixture is flush mounted the Ceiling Cavity height is "0". Dimensions are to be in feet and decimal equivalents of a foot, not in inches (see Step 2 and Table C12-1).

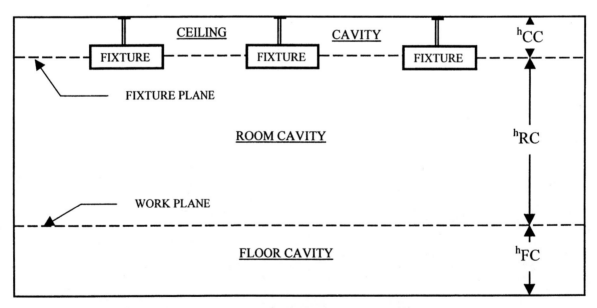

FIGURE C12-7

(b) Determine the Ceiling Cavity Ratio (CCR), the Room Cavity Ratio (RCR), and the Floor Cavity Ratio (FCR), using the following formulas:

$$CCR = \frac{5 \times {}^h CC (L + W)}{L \times W}$$

$$RCR = \frac{5 \times {}^h RC (L + W)}{L \times W}$$

$$FCR = \frac{5 \times {}^h FC (L + W)}{L \times W}$$

Where: CR = Cavity Ratio
                h = height of cavity
                L = Room Length (Step 2)
                W = Room Width (Step 2)

(10) Using the Reflectance percent of Ceiling and Wall (OVHD. and BHD.) from Step 4(b), and the Ceiling or Floor Cavity Ratio (CCR or FCR) from Step 9; fill in the Effective Ceiling Reflectance from Table C12-8. A certain amount of interpolation may be required.

(11) (a) Fill in the Wall Reflectance Percent from Step 4(b), the Room Cavity Ratio (RCR) from Step 9(b), and the Effective Ceiling Reflectance percent from Step 10.

(b) Now the "Co-Efficient of Utilization" (CU) can be determined. This factor should preferably be taken from vendors charts for the specific fixture selected. As an alternative, if the fixture has not been selected, or if the vendors charts are not available, Tables C12-9, 10, and 11 are included as part of this lesson to aid in selecting Incandescent or Fluorescent fixtures.

(12) You now have all the data necessary to fill in the blanks of the formula. This formula will allow a determination of the number of fixtures required for each space.

TABLE C12-8 Effective Ceiling Reflectances

% Ceiling or Floor Reflectance		90				80				70			50			30			
% Wall Reflectance		90	70	50	30	80	70	50	30	70	50	30	70	50	30	65	50	30	10
	0	90	90	90	90	80	80	80	80	70	70	70	50	50	50	30	30	30	30
	0.1	90	89	88	87	79	79	78	78	69	69	68	49	49	48	30	30	29	29
	0.2	89	88	86	85	79	78	77	76	68	67	66	49	48	47	30	29	29	28
	0.3	89	87	85	83	78	77	75	74	68	66	64	49	47	46	30	29	28	27
	0.4	88	86	83	81	78	76	74	72	67	65	63	48	46	45	30	29	27	26
	0.5	88	85	81	78	77	75	73	70	66	64	61	48	46	44	29	28	27	25
	0.6	88	84	80	76	77	75	71	68	65	62	59	47	45	43	29	28	26	25
	0.7	88	83	78	74	76	74	70	66	65	61	58	47	44	42	29	28	26	24
	0.8	87	82	77	73	75	73	69	65	64	60	56	47	43	41	29	27	25	23
	0.9	87	81	76	71	75	72	68	63	63	59	55	46	43	40	29	27	25	22
	1.0	86	80	74	69	74	71	66	61	63	58	53	46	42	39	29	27	24	22
	1.1	86	79	73	67	74	71	65	60	62	57	52	46	41	38	29	26	24	21
	1.2	86	78	72	65	73	70	64	58	61	56	50	45	41	37	29	26	23	20
	1.3	85	78	70	64	73	69	63	57	61	55	49	45	40	36	29	26	23	20
	1.4	85	77	69	62	72	68	62	55	60	54	48	45	40	35	28	26	22	19
	1.5	85	76	68	61	72	68	61	54	59	53	47	44	39	34	28	25	22	18
	1.6	85	75	66	59	71	67	60	53	59	52	45	44	39	33	28	25	21	18
	1.7	84	74	65	58	71	66	59	52	58	51	44	44	38	32	28	25	21	17
	1.8	84	73	64	56	70	65	58	50	57	50	43	43	37	32	28	25	21	17
	1.9	84	73	63	55	70	65	57	49	57	49	42	43	37	31	28	25	20	16
	2.0	83	72	62	53	69	64	56	48	56	48	41	43	37	30	28	24	20	16
	2.1	83	71	61	52	69	63	55	47	56	47	40	43	36	29	28	24	20	16
	2.2	83	70	60	51	68	63	54	45	55	46	39	42	36	29	28	24	19	15
	2.3	83	69	59	50	68	62	53	44	54	46	38	42	35	28	28	24	19	15
	2.4	82	68	58	48	67	61	52	43	54	45	37	42	35	27	28	24	19	14
	2.5	82	68	57	47	67	61	51	42	53	44	36	41	34	27	27	23	18	14
	2.6	82	67	56	46	66	60	50	41	53	43	35	41	34	26	27	23	18	13
	2.7	82	66	55	45	66	60	49	40	52	43	34	41`	33	26	27	23	18	13
	2.8	81	66	54	44	66	59	48	39	52	42	33	41	33	25	27	23	18	13
	2.9	81	65	53	43	65	58	48	38	51	41	33	40	33	25	27	23	17	12
	3.0	81	64	52	42	65	58	47	38	51	40	32	40	32	24	27	22	17	12
	3.1	80	64	51	41	64	57	46	37	50	40	31	40	32	24	27	22	17	12
	3.2	80	63	50	40	64	57	45	36	50	39	30	40	31	23	27	22	16	11
	3.3	80	62	49	39	64	56	44	35	49	39	30	39	31	23	27	22	16	11
	3.4	80	62	48	38	63	56	44	34	49	38	29	39	31	22	27	22	16	11
	3.5	79	61	48	37	63	55	43	33	48	38	29	39	30	22	26	22	16	11
	3.6	79	60	47	36	62	54	42	33	48	37	28	39	30	21	26	21	15	10
	3.7	79	60	46	35	62	54	42	32	48	37	27	38	30	21	26	21	15	10
	3.8	79	59	45	35	62	53	41	31	47	36	27	38	29	21	26	21	15	10
	3.9	78	59	45	34	61	53	40	30	47	36	26	38	29	20	26	21	15	10
	4.0	78	58	44	33	61	52	40	30	46	35	26	38	29	20	26	21	15	9
	4.1	78	57	43	32	60	52	39	29	46	35	25	37	28	20	26	21	14	9
	4.2	78	57	43	32	60	51	39	29	46	34	25	37	28	19	26	20	14	9
	4.3	78	56	42	31	60	51	38	28	45	34	25	37	28	19	26	20	14	9
	4.4	77	56	41	30	59	51	38	28	45	34	24	37	27	19	26	20	14	8
	4.5	77	55	41	30	59	50	37	27	45	33	24	37	27	19	25	20	14	8
	4.6	77	55	40	29	59	50	37	26	44	33	24	36	27	18	25	20	14	8
	4.7	77	54	40	29	58	49	36	26	44	33	23	36	26	18	25	20	13	8
	4.8	76	54	39	28	58	49	36	25	44	32	23	36	26	18	25	19	13	8
	4.9	76	53	38	28	58	49	35	25	44	32	23	36	26	18	25	19	13	7
	5.0	76	53	38	27	57	48	35	25	43	32	22	36	26	17	25	19	13	7

Ceiling or Floor Cavity Ratio (leftmost vertical label)

TABLE C12-9 Photometric Data – Incandescent Fixtures With Globes

Incandescent Fixtures With Globe,
and with or without a guard.

Coefficient of Utilization
Effective Floor Cavity Reflectance 20%

% Reflectance		Room Cavity Ratio									
Eff.Ceil.	Wall	1	2	3	4	5	6	7	8	9	10
80	50	.654	.549	.471	.407	.353	.315	.279	.249	.226	.201
	30	.608	.486	.400	.333	.282	.244	.210	.183	.162	.140
	10	.567	.432	.345	.277	.228	.193	.164	.137	.118	.100
70	50	.603	.506	.434	.375	.326	.289	.257	.230	.208	.185
	30	.563	.449	.370	.309	.260	.225	.194	.170	.150	.130
	10	.523	.401	.320	.257	.211	.177	.150	.127	.110	.093
50	50	.509	.424	.363	.312	.273	.242	.215	.192	.175	.156
	30	.477	.380	.313	.260	.220	.189	.163	.143	.126	.109
	10	.449	.341	.271	.218	.179	.150	.126	.106	.092	.077
30	50	.422	.349	.297	.255	.223	.198	.175	.158	.143	.128
	30	.399	.315	.259	.214	.180	.155	.134	.116	.103	.088
	10	.376	.285	.225	.180	.147	.123	.102	.086	.075	.061
10	50	.343	.280	.237	.202	.176	.157	.140	.124	.114	.101
	30	.324	.253	.206	.170	.142	.122	.106	.091	.080	.068
	10	.307	.230	.180	.143	.115	.096	.080	.066	.057	.045

TABLE C12-10 Photometric Data – Incandescent Fixtures Without Globes

Incandescent Fixtures With Globe, Dome Reflector,
and with or without a guard.

Coefficient of Utilization
Effective Floor Cavity Reflectance 20%

% Reflectance		Room Cavity Ratio									
Eff.Ceil.	Wall	1	2	3	4	5	6	7	8	9	10
80	50	.604	.522	.454	.395	.347	.309	.274	.244	.222	.193
	30	.576	.480	.404	.340	.290	.253	.220	.192	.170	.142
	10	.552	.444	.364	.298	.248	.214	.184	.155	.135	.110
70	50	.591	.512	.445	.387	.339	.303	.270	.241	.219	.189
	30	.565	.472	.398	.336	.286	.250	.217	.191	.168	.142
	10	.543	.440	.361	.296	.246	.211	.181	.154	.135	.110
50	50	.564	.490	.428	.372	.327	.291	.260	.232	.211	.184
	30	.544	.457	.387	.327	.280	.245	.212	.186	.166	.139
	10	.526	.428	.354	.292	.245	.209	.179	.153	.134	.109
30	50	.541	.470	.411	.358	.315	.281	.250	.225	.204	.178
	30	.525	.444	.377	.319	.273	.239	.209	.182	.162	.136
	10	.509	.419	.348	.288	.241	.207	.177	.152	.133	.107
10	50	.521	.452	.396	.345	.304	.272	.243	.217	.198	.173
	30	.507	.429	.367	.311	.268	.234	.205	.179	.159	.134
	10	.494	.409	.342	.283	.238	.205	.175	.150	.131	.106

TABLE C12-11 Photometric Data – Fluorescent Fixtures

Coefficient of Utilization Effective Floor Cavity Reflectance 20%											
2' x 4' – 4 Lamp Fluorescent Fixtures											
% Reflectance		**Room Cavity Ratio**									
Eff.Ceil.	**Wall**	**1**	**2**	**3**	**4**	**5**	**6**	**7**	**8**	**9**	**10**
80	50	.69	.63	.57	.53	.48	.44	.41	.38	.35	.33
	30	.67	.59	.53	.48	.43	.40	.36	.33	.31	.29
50	50	.65	.59	.54	.50	.46	.42	.39	.37	.34	.32
	30	.64	.56	.51	.46	.42	.38	.36	.33	.30	.28
2' x 4' – 2 Lamp Fluorescent Fixtures											
% Reflectance		**Room Cavity Ratio**									
Eff.Ceil.	**Wall**	**1**	**2**	**3**	**4**	**5**	**6**	**7**	**8**	**9**	**10**
80	50	.64	.58	.52	.48	.44	.40	.37	.35	.33	.30
	30	.62	.55	.48	.44	.39	.36	.33	.30	.28	.26
50	50	.58	.53	.48	.45	.41	.38	.36	.33	.31	.29
	30	.57	.51	.46	.41	.37	.34	.32	.30	.27	.25

ILLUMINATION CALCULATION WORKSHEET C12-12

(ZONAL CAVITY METHOD)

1   Space Designation: _____

     Deck: _____ Frames: _____ Side: _____

2   Room Size: _____ Feet Wide X _____ Feet Long X

3   _____ Feet High (Ceiling). Deck Height _____ Feet.

     Room Area: _____ Square Feet.

4   (a) Room Finish: (b) Reflectance percents:

       Overhead _____   Overhead _____

       Bulkhead _____   Bulkhead _____

       Deck _____   Deck _____

5   Fixture Type: No. of Lamps _____, Lamp Wattage _____, Size _____,

       Fluor or Incand. _____, Mounting _____,

6   Reflector _____, Mfgrs. Name & Part No. _____

     Fixture Lumen Output: _____

7   Maintenance Factor (MF): _____

8   Foot Candles Required (FC): _____

9   (a) Cavity Heights   (b) Cavity Ratios:

       Ceiling (^hCC) _____Feet   Ceiling (CCR) _____

       Room (^hRC) _____Feet   Room (RCR) _____

       Floor (^hFC) _____Feet   Floor (FCR) _____

10   Effective Ceiling Reflectance percent: _____

11   (a) Wall Reflectance Percent _____.   (b) Co-efficient of Utilization _____

     Room Cavity Ratio (RCR) _____

     Effective Ceiling Reflectance Percent _____

12   FORMULA: No. Fixt. Req'd. = (F.C. X Area) ÷(Lumens X C.U. X M.F.)

          No. = (_____ X _____) ÷(_____ X _____ X _____)

          No. = (_____) ÷(_____)

          No. = ☐

ILLUMINATION CALCULATION WORKSHEET C12-12

(ZONAL CAVITY METHOD)

① Space Designation: ___*PAINT LOCKER*___

Deck: ___*MAIN*___   Frames: ___*0 - 5*___   Side: ___*CENTERLINE*___

② Room Size: ___*23*___ Feet Wide X ___*25*___ Feet Long X

③ ___*9.5*___ Feet High (Ceiling). Deck Height ___*9.5*___ Feet.

Room Area: ___*575*___ Square Feet.

④ (a) Room Finish: (b) Reflectance percents:
   Overhead ___*WHITE*___   Overhead ___*75%*___

   Bulkhead ___*WHITE*___   Bulkhead ___*75%*___

   Deck ___*DARK GRAY*___   Deck ___*30%*___

⑤ Fixture Type: No. of Lamps ___*1*___, Lamp Wattage ___*100*___, Size ___*4 ¼"DIA.*___,

   Fluor or Incand. ___*INCAND.*___, Mounting ___*SURFACE*___,

⑥ Reflector ___*NONE*___, Mfgrs. Name & Part No. _____

⑦ Fixture Lumen Output: ___*1670*___

Maintenance Factor (MF): ___*0.7*___

⑧ Foot Candles Required (FC): ___*15*___

⑨ (a) Cavity Heights   (b) Cavity Ratios:
   Ceiling (^{h}CC) ___*1.0*___ Feet   Ceiling (CCR) ___*.42*___

   Room (^{h}RC) ___*6.0*___ Feet   Room (RCR) ___*2.50*___

⑩ Floor (^{h}FC) ___*2.5*___ Feet   Floor (FCR) ___*1.04*___

Effective Ceiling Reflectance Percent: ___*67*___

⑪ (a) Wall Reflectance Percent ___*75*___.   (b) Co-efficient of Utilization ___*.47*___

Room Cavity Ratio (RCR) ___*2.5*___

Effective Ceiling Reflectance Percent ___*67*___

⑫ FORMULA: No. Fixt. Req'd. = (F.C. X Area) ÷(Lumens X C.U. X M.F.)

No. = (___*15*___ X ___*575*___) ÷(___*1670*___ X ___*.47*___ X ___*0.7*___)

No. = (___*8625*___)÷(___*549.43*___)

No. = [*16*]

ILLUMINATION CALCULATION WORKSHEET C12-12

(ZONAL CAVITY METHOD)

(1) Space Designation: ___CONTROL ROOM___

Deck: ___UPPER ENG. RM.___ Frames: ___21-22 3/4___ Side: ___STBD___

(2) Room Size: ___14___ Feet Wide X ___20.75___ Feet Long X

(3) ___9.5___ Feet High (Ceiling). Deck Height ___10.5___ Feet.

(4) Room Area: ___290.5___ Square Feet.

(a) Room Finish: (b) Reflectance percents:
Overhead ___LIGHT GRAY___   Overhead ___75%___

Bulkhead ___LIGHT GRAY___   Bulkhead ___75%___

Deck ___DARK GREEN___   Deck ___10%___

(5) Fixture Type: No. of Lamps ___4___, Lamp Wattage ___40___, Size ___4' LONG___,

Fluor or Incand. ___FLUOR.___, Mounting ___FLUSH___,

(6) Reflector ___SHALLOW BOWL___, Mfgrs. Name & Part No. _____

(7) Fixture Lumen Output: ___3150 X 4 = 12,600___

Maintenance Factor (MF): ___0.85___

(8) Foot Candles Required (FC): ___50___

(9) (a) Cavity Heights   (b) Cavity Ratios:
Ceiling (h_{CC}) ___0.0___ Feet   Ceiling (CCR) ___0.00___

Room (h_{RC}) ___7.0___ Feet   Room (RCR) ___4.19___

(10) Floor (h_{FC}) ___2.5___ Feet   Floor (FCR) ___1.50___

Effective Ceiling Reflectance Percent: ___59%___

(11) (a) Wall Reflectance Percent ___75___.   (b) Co-efficient of Utilization ___.51___

Room Cavity Ratio (RCR) ___4.19___

Effective Ceiling Reflectance Percent ___59%___

(12) FORMULA: No. Fixt. Req'd. = (F.C. X Area) ÷(Lumens X C.U. X M.F.)

No. = (___50___ X ___290.5___) ÷(___12,600___ X ___.51___ X ___0.85___)

No. = (___14,525___) ÷(___5462.1___)

No. = [3]

I.   On marine installations, some systems require the use of D.C. Battery Power for operations. These may include, but are not limited to: General Emergency Alarm Systems, Diesel Generator Starting, Navigation Lights, some Radio Installations, etc.

II.   Because most larger vessels have an emergency generator which supplies power for the general alarm system, the current regulations do not require General Alarm batteries [USCG Subpart 113.25-6, SOLAS 74 Regulations II-1/42 or II-1/43]. The regulations, however, still have requirements if batteries are used.

III.   Due to many existing vessels having battery operated General Alarm Systems, the designer or installer may be called upon to order and replace batteries for an existing system, or order batteries for new electronics. This lesson is designed to familiarize you with a method of sizing and ordering batteries and battery chargers.

IV.   Battery manufacturers, in addition to the battery physical characteristics, list the electrical properties of each battery cell in their line.

   A. These characteristics are usually listed as:

      1. Battery type (Nickel Cadmium Alkaline, Lead Acid, or Sealed-Gelled).
      2. Nominal Amp Hour (AH) rating for each cell based on a nominal number of charging hours.
      3. Individual Amps per Hour output.

   B. Batteries are usually ordered based on their <u>Nominal</u> Amp Hours at a <u>Nominal</u> hourly rate for each cell, and the number of cells required to make up the battery.

      Example: 24 volts D.C. – 95 AH @ 10 Hour Rate – 20 Cells

   C. Nameplates are required on batteries with the manufacturers name and battery type designation, rated voltage, amp-hour rating at a specific discharge rate, specific gravity of the electrolyte (for lead acid the specific gravity when fully charged [USCG Subpart 111.15-5(e), ABS Rule 4/5C3.3.3 and Table 4/5C4b, and IEEE Std.45-1998 Sub-clause 6.2.1].

   D. To arrive at the nominal battery rating it is necessary to:

      1. Determine the full load amps of the system or equipment which will be served by the battery
      2. Determine how many charging hours the battery will be required to completely charge the battery.

V.   The proper method for sizing and ordering a battery is as follows:

   A. Determine the voltage of the system or equipment which will be powered by the battery;

Lesson 13

B. Determine the battery type; (Nickel Cadmium, Lead Acid, or Gel-Cell);

C. Determine the volts per cell;

- Nickel Cadmium has 1.2 to 1.4 volts per cell
- Lead Acid has 2.0 volts per cell

D. Determine the number of cells required by dividing the required voltage by the volts per cell:

NO. OF CELLS = REQUIRED VOLTAGE ÷ VOLTS PER CELL

E. Determine the total amp load for the system. Each cell is rated for this amount. No matter how many cells are used the amp rating for the battery is the same as the amp rating for one (1) cell. Adding cells increases the voltage, not the amps.

F. Determine the required charging rate in hours. Eight hours of continuous operation used to be required by USCG until recently. When resizing for an old system assume this 8 hour period. For new systems the capacity should be enough for 30 minutes, when the prime source of power is an emergency generator [IEEE Std. 45-1998 Sub-clause26.2.2].

G. Select a battery cell with all the characteristics above and list the battery voltage, nominal AH at a nominal hourly rate, and the number of cells required.

H. Select a battery charger which will keep the battery fully charged by a "trickle" or "float" charge when the battery is idle, and a "hi-rate" charge when the battery is in use.

- NICAD "Float" Charging = 1.40 - 1.42 Volts
- NICAD "Hi-Rate" Charging = 1.50 - 1.65 Volts

- Lead Acid "Float" Charging = 2.17 Volts
- Lead Acid "Hi-Rate" Charging = 2.33 Volts

VI. <u>EXAMPLE</u>

Question: A General Emergency Alarm System has been in place for some time but any data on the battery has long since been lost. This system is 24 Volts D.C. and has twenty-one (21) ten inch bells, thirty-one (31) eight inch bells, and three (3) rotating beacons. What size battery and charger is required?

SOLUTION:

A. A blank "BATTERY CALCULATION WORKSHEET" C13-3 is provided with this lesson to aid the designer or installer in ordering batteries and battery chargers. A separate worksheet should be made for each set of batteries and chargers required on the vessel.

B. An example of a completed worksheet for this example has been included as the final page of this lesson to show the sizing of the batteries and charger for this General Emergency Alarm System.

| 1 | List the purpose of the battery.

| 2 | List the voltage of the system

| 3 | Determine which type of battery will be used if not listed in the ships specifications, Nickel Cadmium Alkaline (NICAD), Lead Acid, or Gel-Cell [IEEE Std. 45-1998 Sub-clause 6.1].

While Nickel Cadmium batteries require more cells per battery than Lead Acid, they are better for Marine Installations from a maintenance and installation standpoint.

(a) Lead Acid requires constant level checks of the electrolyte while NICAD does not.

(b) Lead Acid battery racks or trays require lead lining, while NICAD only requires steel or a non-corrosive material [USCG Subpart 111.15-5(f)(2) and (3)and IEEE Std.45-1998 Sub-clause 6.3.5(a)].

(c) Lead Acid gives off more fumes than NICAD.

(d) Sealed-gelled electrolyte batteries minimize the quantity of gas released through a pressure relief valve by recombining the products of electrolysis. The electrolyte in this type of battery cannot be replaced [IEEE 45 Std. 45-1998 Sub-clause 6.1].

| 4 | List the nominal volts per cell"

- NICAD = 1.2 Volts per cell
- Lead Acid = 2.0 Volts per cell

5   List the charging volts per cell:

- NICAD "Float" Charging = 1.40 - 1.42 Volts
- NICAD "Hi-Rate" Charging = 1.50 - 1.65 Volts

6   Determine the number of cells required:

NO. OF CELLS = SYSTEM VOLTAGE ÷ VOLTS PER CELL
NO. OF CELLS = 24 ÷ 1.2
NO. OF CELLS = 20

7   Using the space provided on the worksheet, determine the total load for the system in amps:

(a) From manufacturers catalogs we find that 8" and 10" general alarm bells draw 0.14 amps each at 24 VDC, and rotating beacons draw 1.75 amps each at 24 VDC.

(b) By simple math the total amps can be added;

0.14 amps x 21 (10" bells) = 2.94
0.14 amps x 31 (8" bells) = 4.34
1.75 amps x 3 (beacons)   = <u>5.25</u>
Total system amps = 12.53

8   Determine the charging time in hours. Except for batteries that normally stand idle for long periods of time, the charging facilities for any battery should completely charge the battery in a maximum of 8 hours without exceeding a safe charging rate (IEEE Std. 45-1998 Sub-clause 6.7].

9   To arrive at the <u>nominal</u> amp hour rating, the manufacturers charts for the selected battery should be consulted. For the purpose of this lesson to allow familiarization in the practice of selecting a battery type, a typical "BATTERY PERFORMANCE DATA TABLE C13-1" is included in this lesson.

(a) Since we have selected a Nickel Cadmium batter which must deliver at least 12.53 Amps each hour with a charging rate of 8 hours, we can consult Table C13-1 and arrive at a cell type which has a <u>nominal</u> rating of "125 A.H @ 10 Hr. Rate"

10   With all the information previously determined we can fill in the battery type and voltage, number of cells, cell type and nominal capacity. This now enables a search of manufacturers to select a battery with <u>at least</u> the characteristics listed. Then the manufacturer and cell type can be added to the worksheet.

> **11**   Now that the battery is selected, the battery charger with sufficient capacity to keep the system batteries charged can be ordered.

(a) For vessels with a continuous supply of A.C. power, a battery must be charged with a continuous trickle charge unless the battery is being discharged. When the battery is discharged it must be automatically charged at a higher rate [USCG Subpart 112.55-10(b)].

(b) A battery charger must be selected with the output capacity to "trickle" or "float" charge a battery when it is not being used, to maintain it in a fully charged condition, and a "hi-rate" charge to automatically charge it when it is being used (see rating in step 5).

(c) Battery chargers are sized according to battery voltage (24V) and at least the full load amps of the system (12.53 amps). The charger must also be of drip-proof construction.

(d) To arrive at a battery charger size the manufacturers charts for the selected battery charger should be consulted. For the purpose of this lesson to allow familiarization in the practice of selecting a charger, a typical "BATTERY CHARGER TYPES TABLE C13-2" is included in this lesson.

1. List the Input voltage, Frequency, and Phase.
2. List the nominal battery capacity in Amp Hours (from step 9).
3. List the Battery Type (from step 3) and the Number of Cells (from step 6).
4. List the D.C. Load in amps (from step 7) and the Hours of Operation (From step 8).
5. List the "Float" and "Hi-Rate" charge (from step 5).
6. List the operation needed (usually automatic).
7. List the enclosure type (drip-proof).
8. With all the information previously determined we can now conduct a search of manufacturers to select a battery charger with <u>at least</u> the characteristics listed. Then the manufacturer and model number can be added to the worksheet.

VII.   ADDITIONAL INFORMATION

A. If a battery is used for Emergency Generator starting, it must be capable of supplying power for three (3) consecutive starts. In addition a second source of energy must be provided for an additional three (3) consecutive starts within 30 minutes unless manual starting can be demonstrated to be effective [ABS Rule 4/5A3.17.2].

TABLE C13-1 NICAD Battery Performance Data

| Nom. Cap.* (Ah) | \multicolumn{8}{c}{Available Amperes to and END VOLTAGE OF **1.14 Volts per Cell**} |

Nom. Cap.* (Ah)	10 hr.	8 hr.	5 hr.	4 hr.	3 hr.	2 hr.	1 ½ hr.	1 hr.
10	0.96	1.18	1.77	2.11	2.58	3.08	3.25	3.80
17	1.63	2.0	3.0	3.6	4.4	5.25	5.5	6.45
30	2.9	3.5	5.3	6.3	7.7	9.25	9.7	11.4
45	4.3	5.3	7.9	9.5	11.6	13.9	14.5	17
69	4.7	7.1	10.6	12.6	15.5	18.5	19.5	23
95	9.1	11.2	16.8	20	24.5	29	31	36
125	12.0	14.7	22	26	32	38.5	40.5	47.5
175	16.8	20.6	31	37	45	54	57	66.5
205	19.7	24	36	43	53	63	66.5	78
230	22	27	40.5	48.5	59	71	75	87
250	24	29.5	44	52.5	64.5	77	81	95
290	28	34	51	61	74.5	89	94	110
320	30,5	37.5	56.5	67.5	82.5	98	104	122
350	33.5	41	62	74	90	108	114	133
415	40	49	73	87.5	107	128	135	158

*Nominal Capacity is measured at the 10-hour rate of discharge to 1.10 Volts per Cell

| Nom. Cap.* (Ah) | \multicolumn{8}{c}{Available Amperes to and END VOLTAGE OF **1.10 Volts per Cell**} |

Nom. Cap.* (Ah)	10 hr.	8 hr.	5 hr.	4 hr.	3 hr.	2 hr.	1 ½ hr.	1 hr.
10	1	1.24	1.93	2.35	2.9	3.5	3.75	4.20
17	1.7	2.11	3.28	4	4.93	6	6.4	7.1
30	3	3.72	5.8	7.1	8.7	10.5	11.3	12.6
45	4.5	5.6	8.7	10.6	13.1	15.8	16.9	18.9
69	6	7.4	11.6	14.1	17.4	21	22.5	25.2
95	9.5	11.8	18.3	22.3	27.6	33.3	35.6	39.9
125	12.4	15.5	24.1	29.4	36.3	43.8	46.9	53
175	17.5	21.7	33.8	41.5	52	64	69	76
205	20.5	25.4	39.6	48.6	61	75	81	89
230	23	28.5	44.4	55	68	84	91	100
250	25	31	48.1	59	74	92	99	109
290	29	36	56	69	86	106	115	126
320	32	39.7	62	76	94	117	126	140
350	35	43.4	68	83	103	128	138	153
415	41.5	51	80	98	122	152	164	181

*Nominal Capacity is measured at the 10-hour rate of discharge to 1.10 Volts per Cell

TABLE C13-2 Typical Battery Charger Types

FILTERED OR UNFILTERED OUTPUTS					
Output Volts - amps	Output Volts - amps	Output* volts - amps	Output volts - amps	Output volts – amps	Output volts – amps
12 – 3	24 – 3	32 – 3	48 – 3	120 – 3	240 – 3
12 – 6	24 – 6	32 – 6	48 – 6	120 – 6	240 – 6
12 – 10	24 – 10	32 – 10	48 – 10	120 – 10	240 – 10
12 – 15	24 – 15	32 – 15	48 – 15	120 – 15	240 – 15
12 – 20	24 – 20	32 – 20	48 – 20	120 – 20	240 – 20
12 – 30	24 – 25	32 – 30	48 – 25	120 – 25	240 – 30
12 – 40	24 – 30	32 – 40	48 – 30	120 – 30	240 – 40
12 – 50	24 – 35	32 – 50	48 – 40	120 – 35	240 – 50
12 – 60	24 – 40	32 – 60	48 – 50	120 – 40	
12 – 75	24 – 50	32 – 75	48 – 60	120 – 50	
12– 100	24 – 60	32 – 100	48 – 75	120 – 60	
	24 – 75	32 – 125	48 – 100	120 – 75	
	24 – 100	32 – 150	48 – 125	120 – 100	
	24 – 125	32 – 200	48 – 150	120 – 125	
	24 – 150	32 – 250	48 – 200	120 – 150	
	24 – 200	32 – 400	48 – 250	120 – 200	
	24 – 250		48 – 300	120 – 250	
	24 – 300		48 – 400	120 – 300	
	24 – 400		48 – 600	120 - 350	
				120 – 400	
				120 – 600	

*32 Volt chargers only available in unfiltered output. All others available with filtered or unfiltered output.

BATTERY CALCULATION WORKSHEET **C13-3**

	BATTERY DATA
1	Battery Purpose: _____
2	System Voltage: _____ Volts D.C.
3	Battery Type: _____
4	Volts per Cell (Nominal): _____ Volts D. C.
5	Volts per Cell (Float or Trickle Charge): _____ Volts D.C.
	Volts per Cell (Hi-Rate Charge): _____ Volts D. C.
6	Number of Cells: _____
7	Total System Amps: _____ (Calculated below)
	_____
	_____
	_____
8	Charging rate to fully charge: _____Hours
9	Amp Hour Rating: _____ AH @ _____ Hour Rate

	BATTERY ORDERING INFORMATION
10	Battery Type: _____ Voltage _____ Volts D. C.
	Number of Cells: _____ Nominal Capacity: _____ AH @ _____ Hour Rate
	Battery Manufacturer: _____ Cell Type: _____

	BATTERY CHARGER ORDERING INFORMATION
11	Input Voltage: _____ Frequency: _____ Cycles per Second Phases: ____
	Battery Nominal Capacity: _____ AH @ _____ Hour rate
	Battery Type: _____ Number of Cells: _____
	D.C. Amp Load: _____ Amps for _____ Hours Duration
	Float Charge: _____ Volts per Cell Hi-Rate Charge: _____ Volts per Cell
	Type of operation: _____ Type of Enclosure: _____
	Charger Manufacturer: _____ Model Number: _____

BATTERY CALCULATION WORKSHEET **C13-3**

BATTERY DATA

1   Battery Purpose: _____ *GENERAL ALARM SYSTEM* _____

2   System Voltage: _____ *24* _____ Volts D.C.

3   Battery Type: _____ *NICKEL CADMIUM*
4   *ALKALINE* _____

5   Volts per Cell (Nominal): _____ *1.2* _____ Volts D. C.

Volts per Cell (Float or Trickle Charge): _____ *1.40-1.42* _____ Volts D.C.

6   Volts per Cell (Hi-Rate Charge): _____ *1.50-1.65* _____ Volts D. C.

7   Number of Cells: _____ *20* _____

Total System Amps: _____ *12.53* _____ (Calculated below)

_____ *10" BELLS - 21 X 0.14 = 2.94* _____

8   _____ *8" BELLS - 31 X 0.14 = 4.32* _____

9   _____ *BEACONS - 3 X 1.75 = 5.25, TOTAL SYSTEM AMPS 12.53* _____

Charging rate to fully charge: _____ *8* _____ Hours

Amp Hour Rating: _____ *125* _____ AH @ _____ *10* _____ Hour Rate

BATTERY ORDERING INFORMATION

10   Battery Type: _____ *NICKEL CADMIUM ALKALINE* _____ Voltage _____ *24* _____ Volts D. C.

Number of Cells: _____ *20* _____ Nominal Capacity: _____ *125* _____ AH @ _____ *10* _____ Hour Rate

Battery Manufacturer: _____ *(TO BE SELECTED)* _____ Cell Type: _____ *(LATER)* _____

BATTERY CHARGER ORDERING INFORMATION

11   Input Voltage: _____ *115* _____ Frequency: _____ *60* _____ Cycles per Second   Phases: *1*

Battery Nominal Capacity: _____ *125* _____ AH @ _____ *10* _____ Hour rate

Battery Type: _____ *NICKEL CADMIUM* _____ Number of Cells: _____ *20* _____

D.C. Amp Load: _____ *12.53* _____ Amps for _____ *8* _____ Hours Duration

Float Charge: _____ *1.40* _____ Volts per Cell   Hi-Rate Charge: _____ *1.50* _____ Volts per Cell

Type of operation: _____ *AUTOMATIC* _____ Type of Enclosure: _____ *DRIP-PROOF* _____

Charger Manufacturer: _____ *(TO BE SELECTED)* _____ Model Number: _____ *(LATER)* _____

I.   Electrical equipment can malfunction, or not operate at all, if there is not enough voltage delivered at the power consuming equipment. Electrical equipment must function between +6% and –10% of rated voltage [USCG Subpart 111.01-17].

II.   Voltage drop is defined as: The loss in voltage in a conductor between the power source and the power consuming equipment.

      A. Losses in voltage are caused by the resistance of the copper conductors in a cable. The longer the cable length, the greater the voltage drop.

      B. This drop can be lessened by increasing the cable size because this increases the cross sectional area of the conductors causing less resistance.

III.   Voltage drop is dependent upon the:

- Power source voltage
- Amps of the equipment served
- Length of the cable
- Number of conductors
- Cable conductor area in circular mils.

IV.   Voltage drop is expressed as a percentage of power source voltage. Under normal conditions the cross sectional area of conductors are to be so determined that the drop in voltage from the main or emergency switchboard bus-bars to <u>any</u> and <u>every</u> point of the installation when the conductors are carrying the maximum current will not exceed 6% of the nominal voltage [ABS Rule 4/5B3.1.3].

      A. When the cable length of a circuit exceeds fifty feet, and the equipment load is close to the rated ampacity of the cable used, then a careful look must be taken at the voltage drop.

      B. The longest run of each size power and lighting cable on a vessel has to be calculated for voltage drop and submitted to regulatory bodies included in a "LIST OF FEEDERS AND MAINS", to prove that allowable voltage drops have not been exceeded [ABS Rule 4/5A1.1.2].

      C. Table C14-1 shows the maximum permissible voltage drops between the source and the equipment.

Lesson 14

TABLE C14-1

Allowable Voltage Drops [from IEEE Std.45-1998 Section11]

TYPE OF CIRCUIT	ALLOWABLE VOLTAGE DROP
Generator to Switchboard Circuits	1%
Storage Batteries to Distribution Point	1%
Circuits between Main and Emergency Switchboards (Bus – Tie Circuits)	2%
Circuits between Main Switchboards	2%
Shore Power Circuits	2%
Motors and Power Circuits	6%
Galley Circuits	6%
Air Heater Circuits	6%
Lighting Circuits	6%

V.   Steps in Calculating Voltage Drops:

   A. Determine the voltage.

   B. Determine Full Load Amps (FLA) or Demand Load Amps (DLA).

   C. Measure the length of the cable (without adding any percentage for waste).

   D. List the cable type and size, and circular mil area of the cable (from cable manufacturers catalog).

   E. Pick one of the formulas from Table C14-2 which best suits the circuit.

      1. When selecting a cable which is over 52,600 circular mils a correction factor (C.F.) must be inserted into the formula. This is determined by system power factors and cable constants. Table C14-3 shows these factors. When dealing with marine installations the 0.80 power factor column in used for motors, and the 1.00 power factor column is used for lighting.

TABLE C14-2 Voltage Drop Formulas

(a) 2 Conductor power or lighting cables

$$VD\% = \frac{12 \times I \times 2L \times 100}{CM \times V}$$

(b) 3 Conductor power or lighting cables
(52,600 Circular mils or less)

$$VD\% = \frac{1.732 \times 12 \times I \times L \times 100}{CM \times V}$$

(c) 3 Conductor power or lighting cables
(Over 52,600 Circular mils

$$VD\% = \frac{1.732 \times 12 \times I \times C.F. \times L \times 100}{CM \times V}$$

Where: 12 = Assumed resistance of copper in Ohms per Circular Mil foot
I = Current in amps
C.F. = Correction factor determined by system power factor (0.8 for motors, 1.0 For lighting) and cable constants
CM = Circular Mil area of cable (from cable catalog)
V = Voltage at source
L = Length of cable

TABLE C14-3 Correction Factors For Cable Calculations

CABLE AWG	POWER FACTOR OF LOAD							
	1.00	0.95	0.90	0.85	0.80	0.75	0.70	0.65
2	1.00	1.01	0.99	0.96	0.92	0.84	0.76	0.68
1	1.00	1.03	1.01	0.98	0.95	0.88	0.80	0.71
0	1.00	1.05	1.04	1.02	0.99	0.93	0.85	0.77
00	1.00	1.07	1.07	1.05	1.03	0.98	0.91	0.84
000	1.00	1.10	1.11	1.10	1.09	1.04	0.98	0.92
0000	1.00	1.13	1.16	1.16	1.15	1.12	1.07	1.01
250M	1.00	1.17	1.21	1.22	1.22	1.20	1.16	1.10
300M	1.00	1.21	1.26	1.28	1.29	1.28	1.25	1.21
350M	1.00	1.24	1.31	1.34	1.36	1.37	1.35	1.31
400M	1.00	1.32	1.39	1.43	1.45	1.46	1.44	1.40

VI.   EXAMPLE NO. 1:

Question: In Figure C14-4 we are supplying a 30 H.P. Bilge Pump motor from a 440 volt Power Panel. The motor draws 40 amps and is fed with a TQNIA-26 cable which is 95 feet long. Is the voltage drop for circuit 1-P412 good?

440 VOLT-3 ∅
PWR. PNL P412

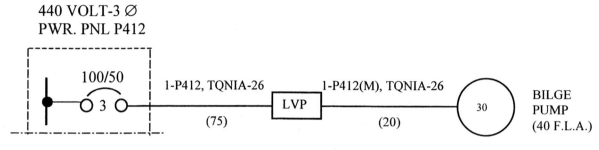

FIGURE C14-4

SOLUTION NO. 1:

A. From manufacturers catalogs find the circular mil area of the cable (26,300 CM).

B. Pick the formula which applies to a 3 conductor cable <u>under</u> 52,600 CM:

$$VD\% = \frac{1.732 \times 12 \times I \times L \times 100}{CM \times V}$$

C. Plug the current, length, circular mils, and voltage into the formula:

$$VD\% = \frac{1.732 \times 12 \times 40 \times 95 \times 100}{26300 \times 440}$$

D. Multiply:

$$VD\% = \frac{7897920}{11572000}$$

E. Divide:

$$VD\% = 0.68$$

F. Since the voltage drop of 0.68 is less than 6%, the cable is good.

VII.   EXAMPLE NO. 2:

Question: From this same panel (Figure C14-5) we are supplying a 150 H.P. Wash Water Pump motor. The motor draws 180 amps and is fed by a TQNIA-300 cable which is 165 feet long. What is the voltage drop of circuit 2-P412?

440 VOLT-3 ∅
PWR. PNL P412

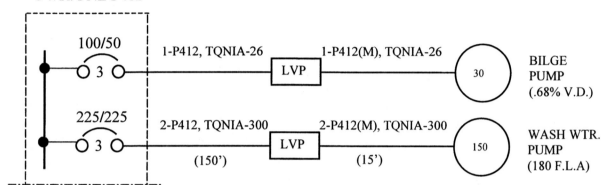

FIGURE C14-5

SOLUTION NO. 2:

A. Find the circular mil area of the cable (300,000 CM).

B. Pick the formula which applies to a 3 conductor cable over 52,600 CM:

$$VD\% = \frac{1.732 \times 12 \times I \times C.F. \times L \times 100}{CM \times V}$$

C. Plug in the values needed
    (for correction factors see step V.E.1 and Table C14-3).

$$VD\% = \frac{1.732 \times 12 \times 180 \times 1.29 \times 165 \times 100}{300000 \times 440}$$

D. Multiply:
$$VD\% = \frac{79629739.2}{132000000}$$

E. Divide:
$$VD\% = 0.60$$

F. Voltage drop is less than 6%, therefore cable is good.

VIII.   EXAMPLE NO. 3:

Question: In Figure C14-6 we are supplying this same panel P412, which has a <u>demand</u> load of 403 amps, with two (2) TQNIA-250 cables which are each 25 feet long. What is the voltage drop of circuit P412 to the panel, and to each piece of equipment from the Main Switchboard?

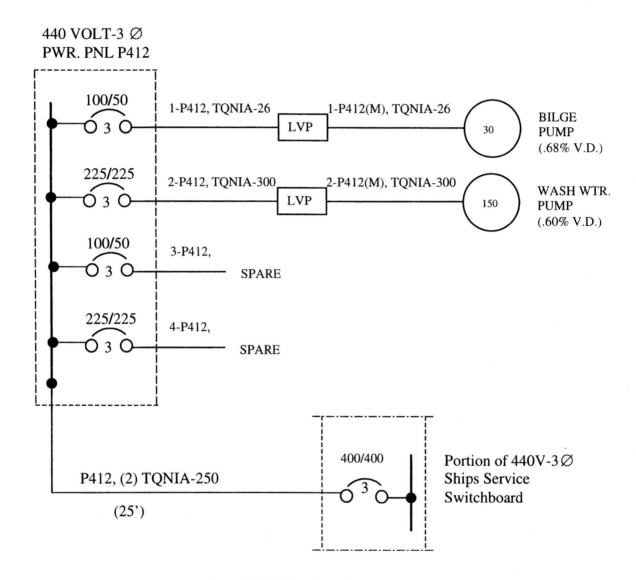

440 VOLT-3 Ø
PWR. PNL P412

FIGURE C14-6

SOLUTION NO. 3:

A. Find the circular mils (250,000 CM).

B. Pick the formula:

$$VD\% = \frac{1.732 \times 12 \times I \times C.F. \times L \times 100}{CM \times V}$$

C. Plug in the values (because each cable will carry half the load, divide the DLA in half, and figure for one cable only. Both will have the same voltage drop).

$$VD\% = \frac{1.732 \times 12 \times 201.50 \times 1.22 \times 25 \times 100}{250,000 \times 440}$$

D. Multiply:

$$VD\% = \frac{12773326.8}{110,000,000}$$

E. Divide:

$$VD\% = 0.11 \text{ (for each cable)}$$

F. We now know that the voltage drop for the feeder P412 is good since it is below 6%.

G. However, since the voltage drops are additive from the power source (Main Swbd.) to the motors, we must add the voltage drop of the Feeder P412 (0.11) to each branch circuit voltage drop. This is to make certain the total voltage drop does not exceed 6%.

1. 1-P412 (0.68)
    +P412 (0.11)
    Total V.D. = 0.79%

2. 2-P412 (0.60)
    +P412 (0.11)
    Total V.D. = 0.71%

H. Thus we can see that all cables for panel P412 are sized correctly.

IX. EXAMPLE NO. 4:

Question: As shown in Figure C14-7, we are supplying a lighting branch circuit of five (5) 100 Watt lights with a DQNIA-4 cable which has a total length of 70 feet. What is the voltage drop of circuit 3-L101?

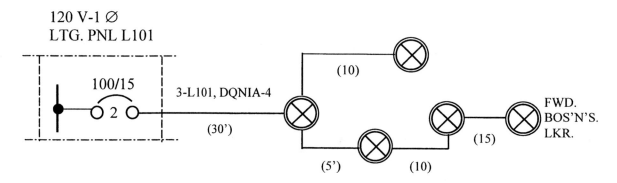

FIGURE C14-7

SOLUTION NO. 4:

A. Instead of using a single length, we have to calculate the Load Center Length
    (LCL) of the circuit (see Figure C14-8). LCL is an average length figured as
    near to the center of the circuit as possible-both load wise and lengthwise.

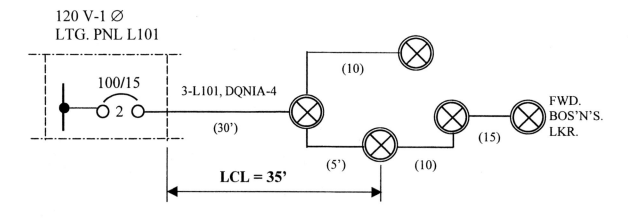

FIGURE C14-8

1. Total Circuit Length: 30 + 10 + 5 + 10 + 15 = 70 feet
2. Load Center Length (LCL):   70 ÷ 2 = 35 feet

B. Pick the formula:

$$VD\% = \frac{12 \times I \times 2L \times 100}{CM \times V}$$

C. Plug in values:

1. Use the load center length of 35 feet instead of a single length.
2. Find the total amps of the entire circuit by:

$$I = P \div E = 500 \div 115 = 4.35 \text{ Amps}$$

$$VD\% = \frac{12 \times 4.35 \times (2 \times 35) \times 100}{4109 \times 115}$$

D. Multiply:

$$VD\% = \frac{365400}{472535}$$

E. Divide:

$$VD\% = 0.77$$

F. Since this is less than 6% we know the cable is good for circuit 3-L101.

X.   VOLTAGE DROP – PROBLEM NO. 1

A. In Figure C14-9 find the voltage drop for each branch circuit and the feeder
   cable. It will be necessary first to size the cables (using "TQNIB" type cables)
   and circuit breakers as learned in Basic Lessons 4 and 5.

440 VOLT-3 ∅
PWR. PNL P419

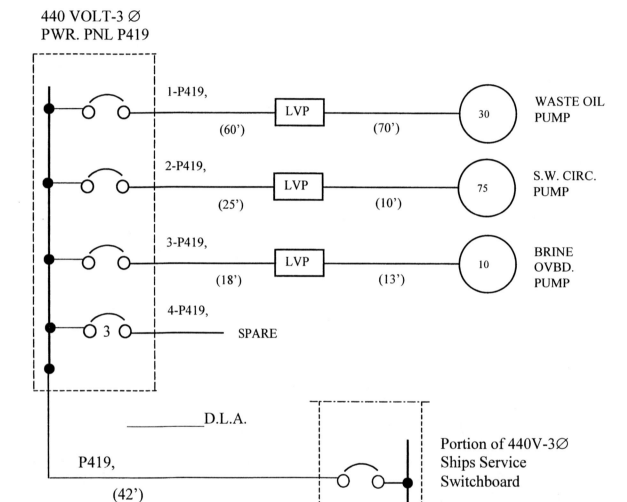

FIGURE C14-9

XI.   VOLTAGE DROP – PROBLEM NO. 2

A. Find the voltage drops for the branch circuits and feeder in Figure C14-10. The light fixture symbol represents Fluorescent Fixtures with two (2) 40 watt lamps in each. Size the cables (using "QNIB" type cables) and breakers per Intermediate Lessons 6 and 7, being mindful of the voltage change between the Main Swbd. and panel L107.

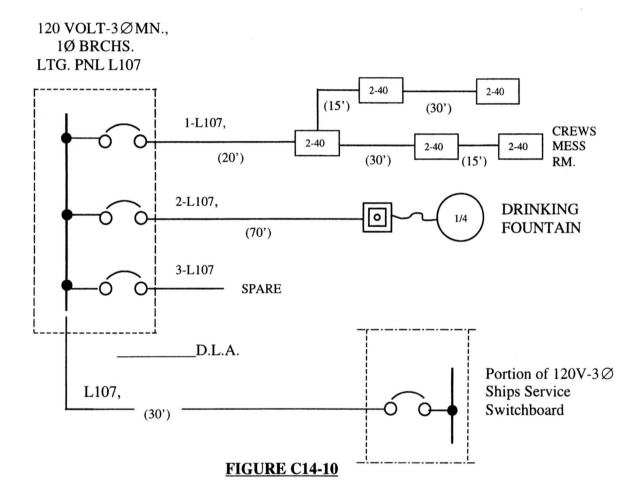

FIGURE C14-10

I.   When designing a new vessel electrical system, a booklet of standard wiring practices and details must be submitted for approval. This booklet must include such items as cable supports, earthing details, bulkhead and deck penetrations, etc. [ABS Rule 4/5B1.1].

II.   Cables and wiring are to be so installed and supported to prevent chafing and other damage. Cables are to be located with a view to avoiding areas of excessive heat and gases, and areas subject to mechanical damage [ABS Rule 4/5B3.1.4 and IEEE Std. 45-1998 Sub-clause 10.3 and 10.4].

III. The installing shipyard needs to know as soon as possible how large to physically construct all wireways, and where to put them. A wireway has to be built before any cables can be installed.

    A. The first step in planning wireways should be to decide on the general route the cables will take. This can best be done by laying out a "Master Wireway Plan" on an isometric background of the ships general arrangements, close to the beginning of the project.

        1. All drawings as they are developed should show cables routed on the ship following the general routing of this "Master Wireway Plan". This should include any "flat" plans for Power and Lighting systems, and isometric drawings of Interior Communications systems.

    B. The second step is to take the drawing schedule for the project and make a Drawing Check List which will list every drawing which will have a cable run on it.

        1. This would include all deck plans and all isometric drawings of Interior Communications systems [USCG Subpart 110.25-1(c) and (d)]. The list would not need to include such drawings as the One-Line Diagram, Load Analysis, List of Feeders and Mains, etc.

    C. Using the "WIREWAY ROUTING WORKSHEET C15-5" included with this lesson we can begin to build the wireway sizes required. Two examples of filled out worksheets C15-5 are provided at the end of this lesson. If used properly, a project will have a multitude of these worksheets which will detail the size and construction of each wireway hanger on the ship.

        1. Using the developed Drawing Check List, pick the first drawing which has cables run on it. Make a separate worksheet for each point where cables pass through a bulkhead, go up or down through a deck, run from switchboards to generators, run to panels, etc.

        2. Select a numbering system for each point selected such as A-A, B-B, A-1, A-2, etc. As each point is selected, a mark should be made on the "Master Wireway Plan" to show where this particular point is located.

Lesson 15

3. Under "Location" fill in the Deck, Frame, Side, and approximate distance off the Centerline.

4. List the designation and size for every cable which passes through each point you have selected.

5. Add the number of the drawing the cables are shown on to the worksheet, then check this number off on the Drawing Check List.

   (a) This procedure is to ensure that no drawings have been left out, or if they are incomplete, their cables can be added later.

6. Proceed through the entire Drawing Check List in this manner. If new points are needed as drawings are developed, it may be necessary to add additional worksheets C15-5.

D. Some points to remember in cable routing:

1. Certain specialty cables which cannot meet the IEEE, ANSI (American National Standards Institute), or IEC flammability requirements, such as certain coaxial cables, cannot be physically run with other cables [USCG 111.60-2(a)].

2. Steering control systems and power units should be separated as widely as practicable from each other [IEEE Std. 45-1998 Sub-clause 18.4.2]. In keeping with the intent of this requirement it is good practice to run the feeder cables to each separate steering gear unit in separate wireways.

3. Some non-shielded signal cables for automation and control systems essential for the safe operation of the vessel should not be run in the same bunch as power or lighting cables, if they are not fiber optic type, or would be subject to electromagnetic interferences [ABS Rule 4/5B3.5.3].

4. No cables will be run through the engine room bulkheads or engine casing unless those cables begin or end in those spaces [ABS Rule 4/5B3.17.1 and IEEE Std. 45-1998 Sub-clause 10.3].

5. No cables shall be run in the galley unless they begin or end in the galley [ABS Rule 4/5B3.17.1 and IEEE Std. 45-1998 Sub-clause 10.3].

6. No cables shall be run in bilges unless they are protected from bilge water [ABS Rule 4/5B3.1.4, USCG Subpart 111.01-5(d), and IEEE Std. 45-1998 Sub-clause 10.3].

7. A clearance of at least the largest cable diameter in a bunch must be maintained between cable bunches [ABS Rule 4/5B3.11.2].

E. After a listing has been made of all cables from all plans, and using the manufacturers catalog for the cables selected, fill in the Outside Diameter (O.D.) and Minimum Bending Radius for each cable.

1. For ABS only classification, the minimum inside bending radius of all cables is six times the cable diameter, except unarmored cables of 1" (25mm) or less diameter. These have a minimum bending radius of four times the cable diameter [ABS Table 4/5B.2].

2. If the U.S.C.G. is involved in the approvals, the minimum bend radius is eight times the cable diameter, except unarmored cables which can be bent to six times the cable diameter [IEEE Std. 45-1998 Sub-clause 10.6].

3. To be on the safe side, and for purposes of this lesson, we will assume the greater radius of eight times the cable diameter for all armored cables, and six time for unarmored.

4. If the minimum bending radius is not given, as for some I.C. cables, this can be computed by multiplying the O.D. of the cable times eight (8) or six (6) as appropriate.

F. Total up the Outside Diameter of all cables which pass through each point. Multiply this total by ten percent to allow for strapping, expansion, etc., and put this total in the box provided.

G. Next circle the largest cable diameter and the largest minimum bend radius listed for each point. Do this for each wireway worksheet.

IV. There should now be enough data collected to decide how large each wireway section has to be, how far apart each tier must be if multiple tiers are used, and what the bending radius of the wireways have to be.

A. Using the "WIREWAY CONSTRUCTION CHART C15-6" included with this lesson, transfer the following cable data from each completed Wireway Routing Worksheet C15-5:

- Wireway point Designation
- Location
- Total Inches Required (Rounded off to next higher full inch)
- Largest Cable O.D.
- Largest Minimum Bend Radius

1. Use one line for each wireway point selected.

B. From the Total Inches Required, select a wireway hanger that will carry <u>at least</u> this much cable. Bear in mind that, depending on the manufacturer of the wireway hangers, some of the hanger length may be taken up for bolting, strapping, etc. Thus an eight (8) inch hanger might be selected that in reality will only hold seven inches of cables.

C. If the total inches required is more than 24 inches it would be best to use smaller hangers which are stacked one above the other to form <u>tiers</u>. This should be decided by the designer as the wireways are being developed, keeping a critical eye on the space limitations, interferences, etc.

D. If more than one tier is used, the clearance between tiers should be at least as much as the largest cable O.D. in the wireway [ABS Rule 4/5B3.11.2]. For instance, if your largest cable O.D. is 4-1/2", all the tiers should be at least 4-1/2" apart. Tiers should never be closer together than 3 inches.

E. An attempt at uniformity should be made when constructing wireways, to keep from having to order small quantities of odd sizes. For instance, if you have ordered mostly eight inch hangers and some of your points only require five inches, go ahead and use eight inch hangers for these. Do not be afraid to allow a little extra room for future expansion.

V.   There should by this point be a Master Wireway Plan with all points roughly located, and a completed Wireway Construction Chart showing the wireway size at each of the selected points.

A. Lay out a "flat" background of the ship for all spaces which will contain a wireway. Because of the amount of information included on a wireway plan, the scale of the drawing should be at least ¼" = 1'-0". This plan may take several sheets to include all the decks required.

B. It is a good idea to now get a wireway plan of some other vessel which has already been done, preferably from the company you are employed by. <u>Look</u> at this plan very carefully to see what method is used to call out and detail the wireway, and do your plan accordingly, but base it on the data you have compiled for your ship.

C. All cable supports are to be spaced <u>less than</u> 24 inches apart in both the vertical and horizontal directions. These supports must have at least a ½ inch cable bearing surface, whether single clips or wireway hangers. Plastic straps are not to be used as cable supports, but may be used <u>in conjunction with</u> metallic straps for securing cables [ABS Rule 4/5B3.9.1].

D. Draw in all the wireways you will need for the ship. Structural plans should be consulted to get the proper stiffener, deck beam, bulkhead sizes, etc., to provide a proper background.

VI.   Check all ventilation and piping plans against the completed wireways. If a pipe or vent duct interferes with a wireway, it may be necessary to make adjustments in the wireway plan to ensure the wireway clears the interference.

    A. Remember that cables are flexible, and wireways and cables in most cases are easier for the shipyard to reroute than pipes or vent ducts.

VII.   COMPENSATING STEEL

    A. Where cables penetrate through watertight bulkheads or decks, stuffing tubes or multi-cable transit devices should be used [ABS Rule 4/5B3.13 and IEEE Std. 45-1998 Sub-clause 10.7]. Consult manufacturers catalogs for these installations.

    B. Where cables penetrate non-watertight bulkheads, decks, beams, stiffeners, etc., holes are cut out large enough for the cables to pass through. If the thickness of the bulkhead or web is less than ¼" (6mm), a chafing bushing must be added [ABS Rule 4/5B3.13 and IEEE Std. 45-1998 Sub-clause 10.7].

    C. The size of holes required should be such that they will not affect the structural strength of the members through which they pass. If the size or position of the hole affects the strength of the structural member, suitable reinforcing of the member should be provided [IEEE Std. 45-1998 Sub-clause 10.10].

    D. In order for the installing activity to properly install the wireways, it is necessary for the designer or installer to determine what size holes to cut, and whether or not the cut-out needs <u>compensating</u> to replace the amount of steel around the cut that has been removed.

        1. For example: A two tier – 8 inch wireway with tier distances of 3" apart has been designed. This wireway must penetrate a non-watertight bulkhead made up of 15.3 pound (#) steel plate.

        2. It is first necessary to determine how much room we need for the cables. Since it is a two tier we need at least six (6) inches of height. Because the hangers are eight (8) inches wide, we need at least a 6" x 8" wide area for the cables.

        3. Due to the possibility of steel stress fractures at corners, there are <u>no square corners</u> on a ship. A wireway cut-out will usually have a radius at the ends of at least half the height (See Figure C15-1).

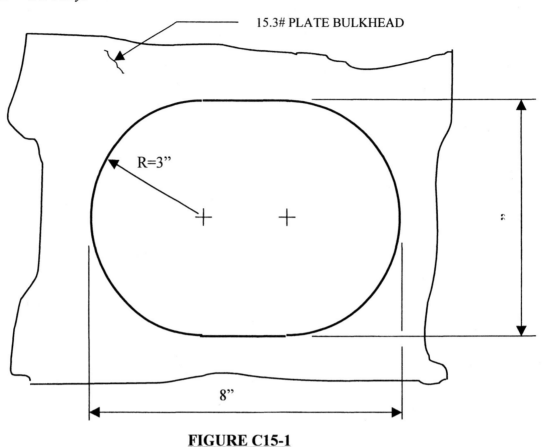

15.3# PLATE BULKHEAD

R=3"

8"

FIGURE C15-1

4. Next find the thickness in inches of the bulkhead which is to be cut. Structural drawings always order steel plate by the pound (#) per square foot, not by the thickness. This can be converted to inches and decimal equivalents of an inch by using TABLE C15-4 WEIGHTS AND MEASUREMENTS OF STEEL PLATES. Since the bulkhead is 15.3# plate, we can see that this is 3/8" (.375") thick.

5. Any bulkhead which is less than ¼" thick has to have a <u>chafing collar</u> added around the hole to protect the cable from chafing damage. Since adding a collar will make the area needed smaller, we should automatically add twice the bulkhead thickness to the area required for cables. We must also do the same if we are compensating for the steel removed by the cut-out.

6. Thus it is necessary to add 3/4" (3/8" thickness x 2) to the Height and Width (6" x 8") of the hole to give a cut-out as shown in Figure C15-2.

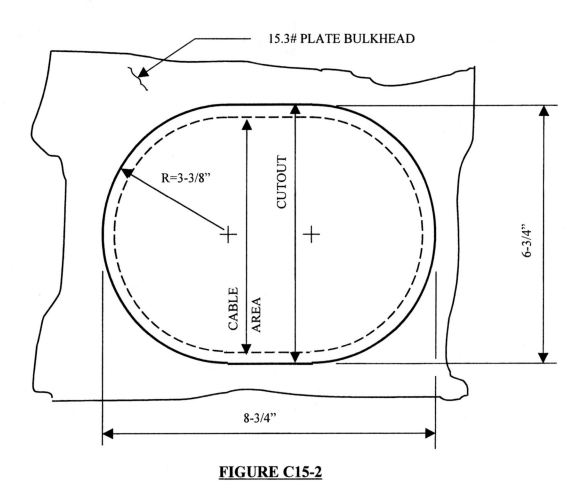

15.3# PLATE BULKHEAD

R=3-3/8"

CUTOUT

CABLE AREA

6-3/4"

8-3/4"

FIGURE C15-2

7. Next figure the amount of steel removed by multiplying the cutout height times the plate thickness at any point along the width (6.75 x .375 = 2.53" of steel removed).

8. This amount of steel has to be compensated for, or <u>put back</u> into the cutout. This is usually done in the form of a "collar" made of steel flat bar (F.B.) which is wrapped around the inside of the cut-out, centered in the bulkhead, and welded in place as shown in Figure C15-3.

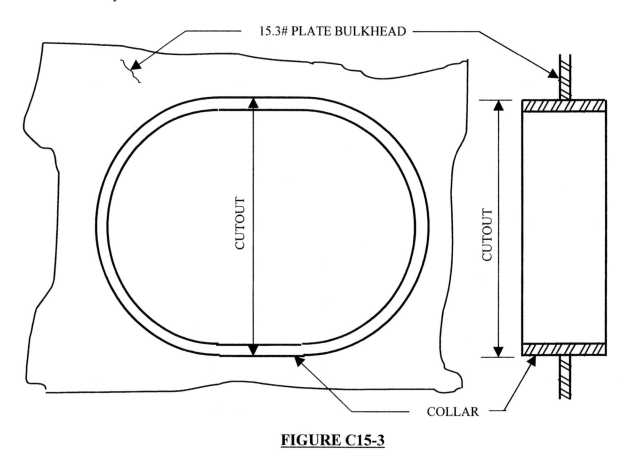

15.3# PLATE BULKHEAD

CUTOUT

CUTOUT

COLLAR

FIGURE C15-3

9. Flat bars come in thicknesses of 1/8" increments, and widths of 1/4" increments. A collar for a wireway is usually never less than 4" wide flat bar.

10. Since the collar is wrapped around the inside of the cut-out and appears at both the bottom and the top, the steel is compensated for if the flat bar collar is at least half as much as the steel removed at any one point.

11. Therefore, if we use a 4" x 3/8" thick flat bar for the collar, we are putting back 1-1/2" (4" x .375" = 1.5") at the top and at the bottom, or a total of 3" (1.5" x 2"). Because this is greater than the 2.53" of steel we have removed by making the cut-out, we have compensated the steel correctly.

TABLE C15-4 Weights And Measurements Of Steel Plates

WEIGHT IN POUNDS PER SQUARE FOOT	THICKNESS IN INCHES		WEIGHT IN POUNDS PER SQUARE FOOT	THICKNESS IN INCHES	
	FRACTION	DECIMAL EQUIVALENT		FRACTION	DECIMAL EQUIVALENT
2.55	1/16	.0625	63.75	1-9/16	1.5625
5.1	1/8	.125	66.3	1-5/8	1.625
7.65	3/16	.1875	68.85	1-11/16	1.6875
10.2	1/4	.25	71.4	1-3/4	1.75
12.75	5/16	.3125	73.95	1-13/16	1.8125
15.3	3/8	.375	76.5	1-7/8	1.875
17.85	7/16	.4375	79.05	1-15/16	1.9375
20.4	1/2	.5	81.6	2	2.00
22.95	9/16	.5625	83.85	2-1/16	2.0625
25.5	5/8	.625	86.7	2-1/8	2.125
28.05	11/16	.6875	89.25	2-3/16	2.1875
30.06	3/4	.75	91.8	2-1/4	2.25
33.15	3/16	.8125	94.35	2-5/16	2.3125
35.7	7/8	.875	96.9	2-3/8	2.375
38.25	15/16	.9375	99.45	2-7/16	2.4375
40.8	1	1.00	102.0	2-1/2	2.5
43.35	1-1/16	1.0625	104.55	2-9/16	2.5625
45.9	1-1/8	1.125	107.1	2-5/8	2.625
48.45	1-3/16	1.1875	109.65	2-11/16	2.6875
51.0	1-1/4	1.25	112.2	2-3/4	2.75
53.55	1-5/16	1.3125	114.75	2-13/16	2.8125
56.1	1-3/8	1.375	117.3	2-7/8	2.875
58.65	1/7/16	1.4375	119.85	2-15/16	2.9375
61.2	1-1/2	1.5	122.4	3	3.00

WIREWAY ROUTING WORKSHEET **C15-5**

JOB NO. _____ SHEET NO. _____ WIREWAY POINT DESIGNATION _____

Location: Deck _____ Frame _____ Side _____ Distance Off Centerline _____

CABLE DESIGNATION	CABLE SIZE	OUTSIDE DIAMETER	MINIMUM RADIUS	DRAWING NUMBER

TOTAL INCHES OF CABLES: _____ X 10% = [   ] TOTAL RE'Q'D.

WIREWAY CONSTRUCTION CHART **C15-6**

JOB NO. _____ SHEET NO. _____

POINT DESIG.	WIREWAY LOCATION DK.-FR.-SIDE	CABLE DATA			WIREWAY DATA		
		TOTAL INCHES	LARGEST O.D.	MIN. BEND RADIUS	SIZE HANGER SELECTED	NO. TIERS REQ'D.	TIER DISTANCE APART

WIREWAY ROUTING WORKSHEET C15-5

JOB NO. _12345_ SHEET NO. _1_ WIREWAY POINT DESIGNATION _A-A (THRU BHD.)_

Location: Deck _SECOND_ Frame _19_ Side _PORT_ Distance Off Centerline _14'-6_

CABLE DESIGNATION	CABLE SIZE	OUTSIDE DIAMETER	MINIMUM RADIUS	DRAWING NUMBER
8-P405	TTNIA-4	.44		S62-1-12
8-P405(M)	TTNIA-4	.44		S62-1-12
8-P105(PB)	TTNIA-4			S62-1-12
3-P404	TTNIA-133	1.55	12.40	S62-1-12
L102	TTNIA-66	1.20		S64-1-1
14-L102	DTNIA-4	.42		S64-1-3
13-L1-2	DTNIA-10	.55		S64-1-3
12-L102	DTNIA-16	.71		S64-1-3
7-L102	DTNIA-10	.55		S64-1-3
2-L102	DTNIA-4	.42		S64-1-3
5-L102	DTNIA-4	.42		S64-1-3
6-L102	DTNIA-4	.42		S64-1-3
3-L1-4	DTNIA-4	.42		S64-1-3
9-L104	DTNIA-4	.42		S64-1-3
10-L104	DTNIA-6	.47		S64-1-3
8-L104	DTNIA-4	.42		S64-1-3
1-L104	DTNIA-4	.42		S64-1-3
C-G11A	DTNIA-4	.42		S65-4-1
C-J7	TP20TIA-10	.77		S65-1-2
C-J15	TP20TIA-10	.77		S65-1-2

TOTAL INCHES OF CABLES: _11.67_ X 1.10% = [12.84] TOTAL RE'Q'D.

WIREWAY ROUTING WORKSHEET C15-5

JOB NO. _12345_   SHEET NO. _2_   WIREWAY POINT DESIGNATION _B-B (THRU DK.)_

Location: Deck _THIRD PLATF._ Frame _14_ Side _STBD_ Distance Off Centerline _6'-2"_

CABLE DESIGNATION	CABLE SIZE	OUTSIDE DIAMETER	MINIMUM RADIUS	DRAWING NUMBER
5-P414	TTNIA-6	.49		S62-1-7
8-EL102	DTNIA-4	.42		S64-1-2
12-L107	DTNIA-4	.42		S64-1-2
2-L107	DTNIA-4	.42		S64-1-2
4-L107	DTNIA-16	.71		S64-1-2
5-L107	DTNIA-10	.55		S64-1-2
6-L107	DTNIA-6	.47		S64-1-2
C-J25	TP20TIA-10	.77	6.16	S65-1-2
C-WD-1	MTNIA-6			S65-5-6
1-EP403	TTNIA-4	.44		S65-5-6

TOTAL INCHES OF CABLES: _5.35_ X1.10% = | 5.89 | TOTAL RE'Q'D.

MARINE ELECTRICAL BASICS

WORKBOOK

APPENDICES

GLOSSARY OF MARINE TERMS

ABOVE BASE LINE (ABL). (See BASE LINE).

ACCOMMODATION LADDER. A portable ladder fastened to a platform attached to the side of the ship and which can be positioned to provide ready access to ship from water or shore.

ACCOMMODATION SPACES. Spaces provided for passengers and crew members that are used for berthing, dining rooms, mess spaces, offices, private baths, toilets and showers, lounges, and similar spaces.

ACROSS-THE-LINE STARTER. A device that connects the motor to the supply without the use of a resistance or autotransformer to reduce the voltage. It may consist of a manually operated switch or a master switch, which energizes and electromagnetically operated contactor.

AFT. Toward, at, or near the stern.

AFTER BODY. That portion of a ship's hull aft of amidships.

AFTER PEAK. The compartment in the stern, abaft the aftermost watertight bulkhead.

AFTER PERPENDICULAR (AP). (See Length Between perpendiculars.)

AIR ESCAPES (AE). Piping leading from a tank or double bottom to a weather deck, so that air may be admitted or discharged in accordance with change of oil or water volume. Also called air pipes, vents, air vents, vapor pipes.

ALTERNATING CURRENT (AC). A periodic current with an average value over a period of time of zero. Unless distinctly specified otherwise, the term refers to a current that reverses at regularly recurring intervals of time and that has alternately positive and negative values.

AMIDSHIPS. In the vicinity of the mid-length of a ship as distinguished from the ends. Technically it is exactly half way between the forward and the after perpendiculars.

ASYNCHRONOUS MACHINE. A machine in which the speed of operation is not proportional to the frequency of the system to which it is connected.

ATHWARTSHIP (ATHW). Across the ship, at right angles to the fore-and-aft centerlines.

AUTOMATIC STARTER. A starter in which the influence directing its performance is automatic.

AUTOTRANSFORMER STARTER. A starter that includes an autotransformer to furnish a reduced voltage for starting a motor. It includes the necessary switching mechanism and is frequently called a compensator or autostarter.

AUXILIARY (AUX) MACHINERY. Various pumps, motors, generators, etc., required on a ship as distinguished from main propulsion machinery units.

BALLAST (BALL). Any solid or liquid weight placed in a ship to increase the draft, to change the trim, or to regulate the stability.

BALLAST TANK. Watertight compartment to hold water ballast.

BASE LINE (BL). A fore-and-aft referenced line at the upper surface of the flat plate keel at the centerline for flush shell plated vessels, or the thickness of the garboard strake above that level for vessels having lap scam shell plating. Vertical dimensions are measured from a horizontal plane through the base line, often called the molded base line.

BEAM, MOLDED. The maximum breadth of the hull measured between the inboard surfaces of the side-shell plating of flush-plated vessels, or between the inboard surfaces of the inside strakes of lap seam-plated vessels.

BERTH. Where a ship is docked or tied up; a place to sleep aboard; a bunk or bed.

BILGE. Curved section between the bottom and the side; the recess into which water drains from holds or other spaces.

BILGE KEEL. A long longitudinal fin fitted on the outside of a ship at the turn of the bilge to reduce rolling. It commonly consists of plating attached to the shell plating by welding or by angles.

BINNACLE. A stand or box for holding and illuminating a compass.

BITT, MOORING. Short metal columns (usually two) extending up from a base plate attached to the deck for the purpose of securing and belaying wire ropes, hawsers, etc., used to secure a vessel to a pier or tug boat; also called a bollard.

BODY PLAN. A drawing consisting of two half transverse elevations or end views of a ship, both having a common vertical centerline, so that the right-hand side represents the ship as seen from ahead, and the left-hand side as seen from astern. On the body plan appear the forms of the various cross sections, the curvature of the deck lines at the side, and the projections, as straight lines of the waterlines, the buttock lines, and the diagonal lines.

BOLTED PLATE MANHOLE (BPMH). A cover for a manhole which is bolted in place.

BOOBY HATCH. An access hatch from a weather deck protected by a hood from sea and weather; also called a companionway.

BOOM. A long round spar hinged at its lower end, usually to a mast, and supported by a wire rope or tackle from aloft to the upper end of the boom. Cargo, stores, etc., are lifted by tackle leading from the upper end of the boom.

BOW. The forward end of a ship.

BRACKET (BKT). A triangular plate used to connect rigidly two or more structural parts such as deck beam to frame or bulkhead stiffener to the deck or tank top.

BREADTH, MOLDED. (See Beam, Molded).

BREAKWATER. A term applied to plates fitted on a forward weather deck to form a V-shaped shield against water that is shipped over the bow.

BRIDGE. A superstructure fitted amidships on the Upper Deck and extending from side to side.

BRIDGE HOUSE. A term applied to an erection fitted amidships on the upper or superstructure deck of a ship. The officers' quarters, staterooms, and accommodations are usually located in the bridge house and the pilot house located above it.

BRIDGE, NAVIGATING OR FLYING. The uppermost deck like platform abreast the pilot house. It generally consists of a narrow walkway supported by stanchions, running from one side of the ship to the other.

BULKHEAD (BHD). A term applied to the vertical partition wall which subdivides the interior of a ship into compartments or rooms. The various types of bulkheads are distinguished by their location, use, kind of material, or method of fabrication, such as fore peak, longitudinal, transverse, watertight, wire mesh, pilaster, etc. Bulkheads which contribute to the strength of a vessel are called strength bulkheads, those which are essential to the watertight subdivision are watertight or oiltight bulkheads, and gastight bulkheads serve to prevent the passage of gas or fumes.

BULWARK. Fore-and-aft vertical plating immediately above the upper edge of the sheerstrake, usually about 3 ft. 6 in. high.

BUNK. A berth or bed, usually built in.

CAMBER. The rise or crown of a deck, athwartship; also called round of beam.

CAPACITANCE (CAPACITY). That property of an system of conductors and dielectrics that permits the storage of electricity when potential differences exist between the conductors. Its value is expressed as the ratio of a quantity of electricity to a potential difference. A capacitance value is always positive.

CAPACITY PLAN. A plan outlining the spaces available for cargo, fuel, fresh water, water ballast, etc., and containing cubic or weight capacity lists for such spaces and a scale showing deadweight capacities at varying drafts and displacements.

CAPSTAN. A revolving device with a vertical axis, used for heaving-in mooring lines.

CARGO PORT. Opening in a ship's side for loading and unloading cargo.

CARGO VESSEL. A vessel that carries bulk, containerized, or roll-on/roll-off dry cargo, and no more than 12 passengers. Research vessels, search and rescue vessels, and tugs are also considered to be cargo vessels by the Institute of Electrical and Electronics Engineers (IEEE).

CASING (CSG), ENGINE AND BOILER.
Bulkheads enclosing a large opening between the weather deck and the engine and boiler rooms. This provides space for the boiler uptakes, access to these rooms, and permits installing or removing large propulsion units such as boilers or turbines. Skylights above casings provide natural ventilation.

CEILINGS. In steel ships, ceilings are wooden planking fitted as flooring on the tank tops in the lower hold. Usually limited to the square of the hatch and used to protect the tank tops in break bulk ships. Also part of joiner installations installed under the steel overheads.

CENTERLINE. The middle line of the ship, extending from stem to stern at any level.

CENTERLINE VERTICAL KEEL (CVK). (See KEEL).

CHAIN LOCKER. Compartment in forward lower portion of ship in which anchor chain is stowed.

CHAMFER. To cut off the sharp edge of a 90-degree corner; to trim to an acute angle.

CHOCK. A heavy smooth surfaced fitting usually located near the edge of the weather deck through which wire ropes or manila hawsers may be led, usually to piers.

CLEAT. A fitting having two arms or horns around which ropes may be made fast; a clip on the frames to hold the cargo battens in place.

CLIP. A short length of angle to attach or connect structural parts.

COAMING, HATCH. The vertical plating bounding a hatch for the purpose of stiffening the edges of the opening and forming the support for the hatch beams.

COFFERDAM (CD). Narrow void space between two bulkheads or floors that prevents leakage between the adjoining compartments.

COMPARTMENTATION (COMP). The subdividing of the hull by transverse watertight bulkheads so that the vessel may remain afloat under certain assumed conditions of flooding.

CONTINUOUS DUTY. A requirement of service that demands operation at a constant load for an indefinite period of time.

COWL. A hood-shaped top or end of a natural ventilation trunk that may be rotated in direction to cause wind to blow air into or out of the trunk.

CROW'S NEST. An elevated lookout station, usually attached to forward side of foremast.

CYCLE. The complete series of values of a periodic quantity that occurs during a period. It is one complete set of positive and negative values of an alternating current.

DAVIT. A crane arm for handling lifeboats, anchors, stores, etc.

DEAD LIGHT OR FIXED LIGHT. A term applied to a port light that does not open.

DEADWEIGHT. The total weight in long tons (2240 lbs.) that a vessel carries on a specified draft including cargo, fuel, water in tanks, stores, baggage, passengers, crew, and their effects but excluding the water in the boilers.

DECK. A deck or platform in a ship corresponding to a floor in a building. It is the plating, planking, or covering of any tier of beams either in the hull or superstructure of a ship. Decks are usually designated by their location as boat deck, bridge deck, upper deck, main deck, etc., but are sometimes designated by letter such as "A" deck, "B" deck, etc.

DECK, FREEBOARD. Deck to which freeboard is measured. Normally the uppermost continuous deck exposed to the weather having permanent means of closing all weather openings.

DECK HEIGHT. The vertical distance between the molded lines of two adjacent decks.

DECKHOUSE. A term applied to a partial superstructure that does not extend from side to side as do the bridge, poop, and forecastle.

DECK MACHINERY. A term applied to steering gear, capstans, windlasses, winches, and miscellaneous machinery located on the decks of a ship.

DEEP TANKS. Tanks extending from the bottom or inner bottom up to or higher than the lowest deck. They are often fitted with hatches so that they also may be used for dry cargo in lieu of fuel oil, ballast water, or liquid cargo.

DEMAND FACTOR. The ratio of the operating load demand of a system or part of a system to the total connected load of the system or part of the system under consideration.

DEPTH, MOLDED. The vertical distance from the molded base line to the top of the freeboard deck beam at side, measured at mid-length of the vessel.

DIRECT CURRENT (DC). A unidirectional current in which the changes in value (polarity) are either zero or so small that they may be neglected. As ordinarily used, the term designates a practically nonpulsating current.

DOG. A small metal fitting used to hold doors, hatch covers, manhole covers, etc., closed.

DOUBLE BOTTOM. Compartments at the bottom of a ship between inner bottom and the shell plating used for ballast water, fresh water, fuel oil, etc.

DRAFT MARKS. The numbers which are placed on each side of a vessel at the bow and stern, and sometimes amidships, to indicate the distance from the lower edge of the number to the bottom of the keel or other fixed reference point. The numbers are 6 in. high and spaced 12 in. bottom to bottom vertically.

DRIPPROOF (DP) ENCLOSURE. An enclosure in which the openings are so constructed that drops of liquid or solid particles falling on the enclosure at any angle not greater than 15^0 from the vertical either cannot enter the enclosure, or if they do enter the enclosure, they will not prevent the successful operation of, or cause damage to, the enclosed equipment (NEMA 2 or 12).

DUNNAGE. Any material, such as blocks, boards, burlap, etc., used in the stowage of cargo.

DUSTPROOF ENCLOSURE. An enclosure so constructed or protected that any accumulation of dust that may occur within the enclosure will not prevent the successful operation of, or cause damage to, the enclosed equipment (NEMA 5).

DUSTIGHT (DT) ENCLOSURE. An enclosure constructed so that dust cannot enter (NEMA 4/4X).

ELECTRIC COUPLING. A device for transmitting torque by means of electromagnetic force in which there is no mechanical torque contact between the driving and driven members. The slip type electric coupling has poles excited by direct current on one rotating member, and an armature winding, usually of the double squirrel cage type, on the other rotating member.

EMBEDDED TEMPERATURE DETECTOR. A resistance thermometer or thermocouple built into a machine for the purpose of measuring the temperature.

ESCAPE TRUNK. A vertical trunk fitted with a ladder to permit personnel to escape if trapped. Usually provided from the after end of the shaft tunnel to topside spaces.

EXCITER. The source of all or part of the field current for the excitation of an electric machine.

EXPANDED METAL BULKHEAD (EMB). A wire mesh bulkhead not used as a supporting member but a separation of spaces. (See also BULKHEAD).

EXPANSION TRUNK OR TANK. A trunk extending above a space which is used for the stowage of liquid cargo. The surface of the cargo liquid is kept sufficiently high in the trunk to permit expansion without risk of excessive strain on the hull or of overflowing, and to allow contraction of the liquid without increase of free surface.

EXPLOSIONPROOF (EP) ENCLOSURE. An enclosure designed and constructed to withstand an explosion of a specified flammable gas or vapor that may occur within it, and to prevent the ignition of flammable gas or vapor in the atmosphere surrounding the enclosure by sparks, flashes, or explosions of the specified gas or vapor that may occur within the enclosure.

FANTAIL. The overhanging stern section of vessels with round or elliptical after endings to upper most decks and which extend well abaft the after perpendicular.

FATHOM. A measure of length, equivalent to 6 linear feet, used for depths of water and lengths of anchor chain.

FATHOMETER. A device to measure the depth of water, by timing the travel of a generated sound wave from the ship to the ocean bottom and return.

FEEDER. A cable or set of conductors that originates at a main distribution center (main switchboard) and supplying secondary distribution centers, transformers, or motor control center. Bus tie circuits between generator and distribution switchboards, including those between main and emergency switchboards, are not considered as feeders.

FIDLEY. The top of engine and boiler room casings on the weather deck; a partially raised deck over the engine and boiler casings, usually around the smokestack.

FLOODABLE LENGTH. The length of vessel which may be flooded without sinking her below her safety or margin line. The floodable length of a vessel varies from point to point throughout her length and is usually greatest amidships and least near the quarter length.

FORE. A term used in indicating portions or that part of a ship at or adjacent to the bow. Also applied to that portion and parts of the ship lying between amidships and the stem; as, forebody, fore hold, and foremast.

FORECASTLE (FOC'S'LE). A superstructure fitted at the extreme forward end of the upper deck.

FOREPEAK. The watertight compartment at the extreme forward end; the forward trimming tank.

FORE AND AFT. In line with the length of the ship; longitudinal.

FORWARD (FWD). In the direction of the stem.

FORWARD OR FORE PERPENDICULAR (FP). (See length between perpendiculars).

FOUNDATION. The structural supports for the boilers, main engines, or turbines and reduction gears are called main foundations. Supports for machinery space auxiliary machinery are auxiliary foundations. Deck machinery supports are called, for example, steering engine foundation, winch foundation, etc.

FRAME (FR). A term used to designate one of the transverse members that make up the riblike part of the skeleton of a ship. The frames act as stiffeners, holding the outside plating in shape and maintaining the transverse form of the ship.

FRAME SPACING. The fore-and-aft distance, heel to heel, of adjacent transverse frames.

FREEBOARD. The distance from the waterline to the upper surface of the freeboard deck at side.

FREEBOARD DECK (See Deck, Freeboard).

FREQUENCY. The number of periods occurring in unit time of a periodic quantity, in which time is the independent variable.

FULL MAGNETIC CONTROLLER. An electric controller having all of its basic functions performed by devices that are operated by electromagnets.

GALLEY. A cook room or kitchen on a ship.

GANGWAY, GANGPLANK. A passageway, side-shell opening, or ladderway used for boarding a ship.

GENERAL-PURPOSE ENCLOSURE. An enclosure that primarily protects against accidental contact and slight indirect splashing but is neither dripproof nor splashproof (NEMA 1).

GIRDER. A continuous member usually running fore and aft under a deck for the purpose of supporting the deck beams and deck. The girder is generally supported by widely spaced pillars.

GIRTH. Any expanded length, such as the length of a frame from gunwale to gunwale.

GROSS TONNAGE. (See tonnage).

GUSSET PLATE. A bracket plate lying in a horizontal, or nearly horizontal, plane.

GYPSY HEAD. A cylinder like fitting on the end of a winch or windlass shaft. Manila line or wire rope is hauled or slacked by winding a few turns around it, the free end being manually held taut as it rotates.

HATCH, HATCHWAY. An opening in a deck through which cargo and stores are loaded or unloaded.

HATCH, BOOBY. An enclosed access hatchway leading from the weather deck to spaces below.

HATCH COAMING. (See COAMING, HATCH).

HAWSE PIPE. Tube through which anchor chain is led overboard from the windlass wildcat on deck through the vessel's side. Bolsters from rounded endings at the deck and shell to avoid sharp edges. Stockless anchors are usually stowed in the hawse pipe.

HEEL. (See LIST).

HERTZ (Hz). The unit of frequency, one cycle per second.

HOLDS. The large spaces below deck for the stowage of cargo; the lowermost cargo compartments.

HULL. The structural body of a ship, including shell plating, framing, decks, bulkheads, etc.

INBOARD (INBD). Inside the ship; toward the centerline.

INDUCTION MACHINE. An asynchronous AC machine that comprises a magnetic circuit interlinked with two electric circuits, or sets of circuits, rotating with respect to each other and in which power is transferred from one circuit to another by electromagnetic induction. Examples of induction machines are induction generators, induction motors, and certain types of frequency converters and phase converters.

INNER BOTTOM (IB). Plating forming the top of the double bottom; also called tank top.

INTERCOSTAL. Made in separate parts; between floors, frames, or beams, etc.; the opposite of continuous.

INTERMITTENT DUTY. A requirement of service that demands operation for alternate periods: (1) load and not load; or (2) load and rest; or (3) load, no load and rest, as specified.

JACK STAFF. A flagstaff at the bow.

JACOB'S LADDER. A ladder having either manila line, wire rope, or chain sides to which rungs are attached at regular intervals. It is used to get from a ships deck to small craft or to a pier.

KEEL. The principal fore-and-aft member of a ship's framing which runs along the centerline of the bottom and connects to the stem and stern frame and to which is attached the floors of the ship.

KEEL, BILGE. (See BILGE KEEL).

KEEL BLOCKS. Heavy wood or concrete blocks on which ship rests during construction.

KING POST (KP). A strong vertical post used instead of a mast to support a boom and rigging to form a derrick; also called samson post.

KNOT. A unit of speed, equaling one nautical mile per hour; the international nautical mile is 1852 meters or 6076.1 ft.

KNUCKLE. An abrupt change in direction of the plating, frames, keel, deck, or other structure of a vessel.

LAP. A joint in which one part overlaps the other.

LENGTH, OVERALL. (LOA). The extreme length of a ship measured from the foremost point of the stem to the aftermost part of the stern.

LENGTH BETWEEN PERPENDICULARS (LBP).. The length of a ship between the forward and after perpendiculars. The forward perpendicular is a vertical line at the intersection for the fore side of the stem and the summer load waterline. The after perpendicular is a vertical line at the intersection of the summer load line and the after side of the rudder post or stern post, or the centerline of the rudder stock if there is no rudder post or stern post. In no case is the length to be taken at less than 96 percent of the length on the summer waterline. This is the length as defined in the International Load Line regulations and Classification Society Rules.

LIGHTENING HOLE. A hole cut in a structural member to reduce its weight.

LIGHTING BRANCH CIRCUIT. A circuit that supplies energy to lighting outlets. A lighting branch circuit may also supply portable desk or bracket fans, small heating appliances, motors of 190 Watts (1/4 hp) and less, and other portable apparatus of not over 600 Watts each.

LIMBER HOLE. A small hole or slot in a frame or plate for the purpose of preventing water or oil from collecting; a drain hole.

LINE (PLAN). The plans that show the shape or form of the ship. From the lines drawn full size on the mold-loft floor, templates are made for the various parts of the hull.

LINE SHAFTING. Sections of the main shafting located in the shaft tunnel between the engine room and the after peak bulkhead.

LIST. If the centerline plane of a vessel is not vertical, as when there is more weight on one side than on the other, she is said to list, or to heel.

LOAD WATERLINE. The line on the lines plan of a ship representing the intersection of the ships form with the plane of the water surface when the vessel is floating at the summer freeboard draft or the designed draft.

LOCKED-ROTOR TORQUE. The minimum torque of a motor developed for all angular positions of the rotor, when at rest, and with rated voltage and frequency applied.

LONGITUDINAL CENTER OF GRAVITY (LCG). Location of the center of gravity on a longitudinal plane. (See METACENTRIC HEIGHT).

LONGITUDINALS (LONG'L) Fore-and-aft structural shape or plate members below decks, flats, or the inner bottom, or on the inboard side of the shell plating.

LORAN. A long-range radio navigational aid of the hyperbolic type whose position lines are determined by the measurement of the difference in the time of arrival of synchronized pulses. These devices are rapidly being replaced by satellite-based global position systems.

MACHINERY CONTROL ROOM. An enclosed or separated space generally located within the machinery spaces that functions as a central control station.

MACHINERY SPACES. Spaces that are primarily used for machinery of any type, or equipment for the control of such machinery, such as boiler, engine, generator, motor, pump, and evaporator rooms.

MANHOLE. A round or oval hole cut in decks, tanks, etc., for the purpose of providing access.

MANIFOLD (MFLD). A point of distribution in a piping system consisting of one pipe input and multiple outputs with or without valves all on a common header.

MANUAL CONTROLLER. An electric controller having all of its basic functions performed by devices that are operated by hand.

MARGIN PLATE. The outboard boundary of the double bottom, connecting the inner bottom to the shell plating at the bilge.

MAST. A tall vertical or raked structure usually of circular section located on the centerline of a ship and used to carry navigation lights, radio antennae, and usually cargo booms (see KING POST).

MESS ROOM. Dining room for officers or crew.

METACENTER. The center of buoyancy of a listed vessel is not on the vertical centerline plane. The intersection of a vertical line drawn through the center of buoyancy of a slightly listed vessel intersects the centerline plane at a point called the metacenter.

METACENTRIC HEIGHT. The distance from the metacenter to the center of gravity of a vessel. If the center of gravity is below the metacenter the vessel is stable.

MIDSHIP. (See AMIDSHIPS).

MIDSHIP SECTION. A drawing showing a typical cross section of the hull and superstructure at or near amidships and giving the scantlings of the principal structural members.

MOLDED-CASE CIRCUIT BREAKER. A circuit breaker assembled as an integral unit in a supporting and enclosing housing of insulating material; the overcurrent and tripping means being of the thermal type, the magnetic type, the electronic type, or a combination thereof.

MOLD LOFT. A floor space used for laying down (laying off) the full-size lines of a vessel and making templates therefrom for laying out the structural work entering into the hull.

MOORING. Securing a ship at a dock or elsewhere by several lines or cables so as to limit her movement.

MOORING RING. A round or oval casting inserted in the bulwark plating through which the mooring lines, or hawsers, are passed.

MOTOR BRANCH CIRCUIT. A branch circuit that supplies energy to one or more motors and associated motor controllers.

MOTOR CONTROL CENTER (MCC). A group of devices assembled for the purpose of switching and protecting a number of load circuits. The control center may contain transformers, contactors, circuit breakers, protective, and other devices intended primarily for energizing or de-energizing load circuits.

MOTOR-GENERATOR SET. A machine that consists of one or more motors mechanically coupled to one or more generators to convert electric power from one frequency to another, or to create an isolated power source.

MULTICABLE PENETRATOR. A device consisting of multiple nonmetallic cable seals assembled in a surrounding metal frame, for insertion in openings in decks, bulkhead, or equipment enclosures and through which cables may be passed to penetrate decks or bulkheads or to enter equipment without impairing their original fire or watertight integrity.

MULTISPEED MOTOR. A motor that can be operated at any one of two or more definite speeds, each being practically independent of the load. For example, a DC motor with two armature windings, or an induction motor with windings capable of various pole groupings.

NAUTICAL MILE. (See KNOT).

NET TONNAGE. (See TONNAGE).

OFFSETS. A term used for the coordinates of a ships's form, deck heights, etc.

OILPROOF (OP) ENCLOSURE. An enclosure constructed so that oil vapors, or free oil not under pressure, that may accumulate within the enclosure will not prevent successful operation of, or cause damage to, the enclosed equipment (NEMA 13).

OILTIGHT (OT) ENCLOSURE. An enclosure constructed so that oil vapors or free oil not under pressure, which may be present in the surrounding atmosphere, cannot enter the enclosure (NEMA 13).

OUTBOARD (OUTBD). Abreast or away from the centerline towards the side; outside the hull.

OVERBOARD (OVBD). Over the side of the ship.

OVERHANG. That portion of a vessel's bow or stern clear of the water which projects beyond the forward and after perpendiculars.

OVERHEAD (OVHD). The steel section above a space; the underside of the deck above a space. (See also CEILINGS).

PAD EYE. A fitting having one or more eyes integral with a base to provide means of securing blocks, wire rope, or manila line.

PASSENGER VESSEL. A vessel that carries more than 12 persons in addition to the crew.

PILLAR. Vertical member or column giving support to a deck girder, flat, or similar structure; also called stanchion.

PLATFORM. A short deck, usually without camber or sheer. Also called a flat.

PLIMSOLL MARK. A freeboard marking on the sides designating the maximum draft to which the vessel may be loaded in different bodies of water during various seasons of the year.

POOP. A superstructure fitted at the after end of the Upper Deck.

PORT, CARGO. An opening in the side plating provided with a watertight cover or door and used for loading and unloading.

PORTHOLE, PORT LIGHT. A hinged circular glass window in the ships side or deckhouse, for light and ventilation. Also called airport, or side scuttle.

PORT SIDE. The left-hand side of a ship when looking forward; opposite to starboard.

POWER INVERTER. A component for converting DC power into AC power.

PROPELLER. A revolving screw-like device that drives the ship through the water, consisting of two or more blades; sometimes called a screw or wheel.

PROPELLER SHAFT. The short aftermost section of the main shafting to which the propeller is attached; also called tail shaft.

QUARTERS. Living or sleeping rooms.

RABBET. A groove, depression, or offset in a member into which the end or edge of another member is fitted, generally so that the two surfaces are flush. A rabbet in the stem or stern frame would take the ends or edges of the shell plating, resulting in a flush surface.

RAISED WATERTIGHT HATCH (RWTH). A hatch which is not flush with the deck and is capable of being closed to ensure watertight integrity. (See also COAMING, HATCH).

RADAR. A device that radiates electromagnetic waves and utilizes the reflection of such waves from distant objects to determine their existence or position.

RAIL. The rounded wooden member at the upper edge of the bulwark, or the horizontal pipes forming a fence-like railing fitted instead of a bulwark.

RAKE. A term applied to the fore-and-aft inclination from the vertical, of the mast, smokestack, stempost, etc.

RECTIFIER. A component for converting AC to DC power by inversion or suppression of alternate half cycles.

RIGGING. Wire ropes, manila line, tackle, etc., used to support masts, spars, booms, etc., and for handling and placing cargo on board.

ROLL. To impart curvature to a plate; also the transverse motion of the ship in waves.

RUDDER. A device used to steer a vessel. The most common type consists of a vertical metal area, hinged at the forward edge to the stern post or rudder post.

SCANTLINGS. The dimensions of a ship's frames, girders, plating, etc.

SCUPPERS. Drains from decks to carry off accumulations of rain water, condensation, or sea water. Scuppers are located in the gutters or waterways, on open decks, and in corners of enclosed decks, and connect to pipes usually leading overboard.

SCUTTLE. A small circular or oval opening fitted in decks to provide access. When used as escape scuttles and fitted with means whereby the covers can be opened quickly to permit exit, they are called quick-acting scuttles.

SCUTTLE BUTT. A container for drinking water; a drinking fountain.

SEA CHEST. An opening for supplying sea water to condensers, pumps, etc., and for discharging water from the ship's water systems to the sea. It is a cast or built-up structure located in the hull below the waterline and having means for the attachment of the associated piping. Suction sea chests are fitted with strainers or gratings and sometimes have a lip that forces water into the sea chest when under way.

SEAM. Fore-and-aft joint of shell plating, deck, and tank-top plating, or a lengthwise edge joint of any plating.

SERIES-WOUND MOTOR. A DC motor in which the field circuit is connected either in parallel with the armature circuit or to a separate source of excitation voltage.

SEMICONDUCTOR RECTIFIER (SCR). A device consisting of a conductor and semiconductor forming a junction. The junction exhibits a difference in resistance to current flow in the two directions through the junction. This results in effective current flow in one direction only. The semiconductor rectifier stack is a single columnar structure of one or more semiconductor rectifier cells.

SEMI-GUARDED. An enclosure in which all of the openings, usually in the top half, are protected as in the case of a "guarded enclosure", but the others are left open (NEMA 1).

SEPARATELY VENTILATED MACHINE. A machine that has its ventilating air supplied by an independent fan or blower external to the machine.

SHAFT HORSEPOWER (SHP). The vessel horsepower measured at the propeller shaft.

SHAFT TUNNEL, SHAFT ALLEY. A watertight enclosure for the propeller shafting large enough to walk in, extending aft from the engine room to provide access and protection to the shafting in way of after cargo holds.

SHAPE. A rolled bar of constant cross section such as an angle, bulb angle, channel, etc.; also to impart curvature to a plate or other member.

SHEER. The longitudinal curve of a vessel's decks in a vertical plane, the usual reference being to the ship's side; in the case of a deck having a camber, its centerline sheer may also be given in offsets. Due to sheer a vessel's deck height above the base line is higher at the ends than amidships.

SHEERSTRAKE. The course of shell plating at strength deck level.

SHELL EXPANSION. A plan showing the seams and butts, thickness, and associated welding of all plates comprising the shell plating, framing, etc.

SHELL PLATE. The plates forming the outer side and bottom skin of the hull.

SHORE. A brace or prop used for support while building a ship, or for securing break-bulk cargo.

SHROUD. One of the principal members of the standing rigging, consisting of wire rope which extends from the mast head to the vessel's side, affording lateral support for a mast.

SHUNT-WOUND MOTOR. A DC motor in which the field circuit is connected either in parallel with the armature circuit or to a separate source of excitation voltage.

SIGHT EDGE. The visible edge of shell plating as seen from outside the hull.

SINGLE-PHASE CIRCUIT. A circuit energized by a single alternating electromotive force. A single-phase circuit is usually supplied through two conductors. The currents in these two conductors, counted outward from the source, differ in phase by 180^0, or a half cycle.

SKEG. A deep vertical finlike projection on the bottom of a vessel near the stern, installed to support the lower edge of the rudder, to support the propeller shaft for single screw ships, and for the support of the vessel in dry dock; also to minimize erratic steering in seaway.

SKIDS. A skeleton framework used to hold structural assemblies above ground to facilitate welding.

SMOKESTACK. A metal chimney or passage through which smoke and combustion gases are led from the boiler uptakes to the open air; also called a funnel.

SOUNDING PIPE (SP, ST). A pipe leading to the bottom of an oil or water tank, used to guide a sounding tape or jointed rod when measuring the depth of liquid in the tank; also called a sounding tube.

SPLASHPROOF ENCLOSURE. An enclosure in which the openings are so constructed that drops of liquid or solid particles falling on the enclosure or coming towards it in a straight line at any angle not greater than 100^0 from the vertical cannot enter the enclosure either directly or by striking and running along a surface (NEMA 3S).

SQUIRREL-CAGE INDUCTION MOTOR. A motor in which the secondary circuit consists of a squirrel-cage winding suitably disposed in slots in the secondary core.

SQUIRREL-CAGE WINDING. A permanently short-circuited winding, usually uninsulated (primarily used in induction machines) having its conductors uniformly distributed around the periphery of the machine and joined by continuous end rings.

STABILITY. The tendency of a ship to remain upright or the ability to return to her normal upright position when listed by the action of waves, wind, etc.

STABILIZED SHUNT-WOUND MOTOR. A shunt-wound motor that has a light series winding added to prevent a rise in speed, or to obtain a slight reduction in speed, with increase of load.

STANCHION (STAN). Vertical columns supporting decks, flats, girders, etc.; also called a pillar. Rail stanchions are vertical metal columns on which fence-like rails are mounted (see RAIL).

STANDING RIGGING. Fixed rigging supporting the masts such as shrouds and stays. Does not include running rigging such as boom topping lifts, vangs, and cargo falls.

STARBOARD SIDE. The right-hand side of a ship when looking forward; opposite to port.

STARTER. An electric controller that is used to accelerate a motor from rest to normal speed and to stop the motor. A device designed for starting a motor in either direction of rotation includes the additional function of reversing and should be designated a controller.

STEERING GEAR. A term applied to the steering wheels, leads, steering engine, and fittings by which the rudder is turned. Usually applied to the steering engine.

STEM. The bow frame forming the apex of the intersection of the forward sides of a ship. It is rigidly connected at its lower end to the keel and may be a heavy flat bar or of rounded plate construction.

STERN. After end of a ship.

STERN FRAME. Large casting, forging, or weldment attached to after end of the keel. Includes the vertical rudder post, propeller post, and aperture for the propeller; called sternpost or rudder post in nonpropelled and multiple screw vessels.

STERN TUBE. The watertight tube enclosing and supporting the propeller shaft. It consists of a cast-iron or cast-steel cylinder fitted with bearing surface upon which the propeller shaft, enclosed in a sleeve, rotates.

STIFFENER. An angle, T-bar, channel, built-up section, etc., used to stiffen plating of a bulkhead, etc.

STRAKE. A course, or row, of shell, deck, bulkhead, or other plating.

STRINGER. A term applied to a fore-and-aft girder running along the side of a ship at the shell and also to the outboard strake of plating on any deck; the side pieces of a ladder or staircase into which the treads and risers are fastened.

STRUT. Outboard column-like support or vee-arranged supports for the propeller shaft, used on some ships with more than one propeller instead of bossings. Rarely used on merchant vessels.

SUBMERSIBLE ENCLOSURE. An enclosure constructed so that the equipment within it will operate successfully when submerged in water under specified conditions of submergence depth and time (NEMA 6P).

SUPERSTRUCTURE. A decked over structure above the upper deck, the outboard sides of which are formed by the shell plating as distinguished from a deckhouse that does not extend outboard to the ship's sides.

SWASH BULKHEAD, SWASH PLATE. Longitudinal or transverse nontight bulkheads fitted in a tank to decrease the swashing action of the liquid contents, as a vessel rolls and pitches at sea. Their function is greatest when the tanks are partially filled. Without them the unrestricted action of the liquid against the sides of the tank might be severe. A plate serving this purpose but not extending to the bottom of the tank is called a swash plate.

SYNCHRONOUS MOTOR. A polyphase AC motor with separately supplied DC field and auxiliary (amortisseur) winding for starting purposes. The operating speed is fixed by the frequency of the system and the number of poles of the motor. Thus the speed of the motor can be varied by varying the frequency of the power source. The synchronous motor generally operates at unity power factor and can be used to improve the system power factor. It is generally the motor of choice for AC propulsion systems.

TANK, SETTLING. Fuel-oil tanks used for separating entrained water from the oil. The oil is allowed to stand for a few hours until the water has settled to the bottom, when the latter is drained or pumped off.

TANK TOP (TT) (See INNER BOTTOM)

TANK VESSEL. A vessel that carries liquid or gaseous cargo in bulk.

TANK, WING. Tanks located well outboard adjacent to the side shell plating, often consisting of a continuation of the double bottom up the sides to a deck or flat.

TELEGRAPH. An apparatus, either electrical or mechanical, for transmitting orders, as from a ship's bridge to the engine room, steering-gear room, or elsewhere about the ship.

TELEMOTOR. A device for operating the control valves of the steering engine from the pilothouse, either by fluid pressure or by electricity.

THREE-PHASE CIRCUIT. A combination of circuits energized by alternating electromotive forces that differ in phase by one-third of a cycle (120⁰).

TONNAGE (GROSS, ETC.). A measure of the internal volume of spaces within a vessel in which 100 cu. Ft. is 1 ton. Gross tons includes a ship's internal volume excluding such spaces as the double bottom, peak or deep tanks used only for water ballast, open-ended poop, bridge, or forecastle, certain light and air spaces, skylights, anchor and steering-gear spaces, the wheelhouse, toilets, and certain passenger spaces. Net tonnage is the gross tonnage less certain additional spaces such as officer and crew spaces, chart room, and a percentage of the propelling machinery spaces.

TONNAGE OPENINGS. Nonwatertight openings in the shelter deck and in the 'tween deck bulkheads immediately below in order to exclude spaces from tonnage measurement and thus obtain reduced gross and net tonnage; also fitted at ends of partial superstructures. The opening may be closed by nonwatertight wood shifting boards or metal covers meeting the Tonnage and Load Line Regulations.

TOPPING LIFT. A wire rope or tackle extending from the head of a boom to a mast, or to the vessel's structure, for the purpose of supporting the weight of the boom and its loads, and permitting the boom to be raised or lowered.

TOTALLY ENCLOSED FAN-COOLED MACHINE (TEFC). A totally enclosed machine equipped for exterior cooling by means of a fan or fans integral with the machine but external to the enclosing parts.

TOTALLY ENCLOSED NON-VENTILATED MACHINE (TENV). A machine enclosed to prevent the free exchange of air between the inside and outside of the case, but not sufficiently enclosed to be airtight.

TOTALLY ENCLOSED WATER/AIR COOLED MACHINE (TEWAC). A totally enclosed machine with integral water-to-air heat exchanger and internal fan to provide close-loop air cooling of the windings.

TRANSOM FRAME. The aftermost transverse side frame, abaft of which are the cant frames.

TRANSVERSE (TRANS). At right angles to the fore-and-aft centerline.

TRIM. The difference in feet between the draft forward and the draft aft. If the draft forward is the greater the vessel is said to "trim by the head". If the draft aft is greater she is "trimming by the stern". To trim a ship is to adjust the location of cargo, fuel, etc., so as to result in the desired drafts forward and aft.

TRIPPING BRACKET. Flat bars or plates fitted at various points on deck girders, stiffeners, or beams as reinforcements to prevent their free flanges from tripping.

TRUNK (TRK). A vertical or inclined space or passage formed by bulkheads or casings, extending one or more deck heights, around openings in the decks, through which access can be obtained and cargo, stores, etc., handled, or ventilation provided without disturbing or interfering with the contents or arrangements of the adjoining spaces.

'TWEEN DECKS. The space between any two adjacent decks.

UNDERVOLTAGE OR LOW-VOLTAGE PROTECTION (LVP). The effect of a device, operative on the reduction or failure of voltage, to cause and maintain the interruption of power in the main circuit.

UNDERVOLTAGE OR LOW-VOLTAGE RELEASE (LVR). The effect of a device, operative on the reduction or failure of voltage, to cause the interruption of power to the main circuit, but not to prevent the re-establishment of the main circuit on return of voltage.

UPTAKE. A metal casing connecting the boiler smoke outlet with the inner smokestack.

VANG. Wire rope or tackle secured to the outer end of a cargo boom, the lower end being secured to the deck, used to swing the boom and hold it in a desired position.

VARYING-SPEED MOTOR. A motor whose speed varies with the load, ordinarily decreasing when the load increases, such as a series-wound or repulsion motor.

VERTICAL CENTER OF GRAVITY (VCG). Location of the center of gravity on a vertical plane. (See METACENTRIC HEIGHT).

VITAL SERVICES. Services normally considered to be essential for the safety of the ship and its passengers and crew. These usually include propulsion, steering, navigation, fire fighting, emergency power, emergency lighting, electronics, and communications functions. The identification of all vital services in a particular vessel is generally specified by the government regulatory agencies.

WATERLINE (WL). The line of the water's edge when the ship is afloat; technically, the intersection of any horizontal plane with the molded form.

WATERPROOF (WP) ENCLOSURE. An enclosure constructed so that any moisture or water leakage that may occur into the enclosure will not interfere with its successful operation. In the case of motor or generator enclosures, leakage that may occur around the shaft may be considered permissible provided it is prevented from entering the oil reservoir and provision is made for automatically draining the motor or generator enclosure (NEMA 4/4X).

WATERTIGHT (WT) ENCLOSURE. An enclosure constructed so that any moisture or water leakage that may occur into the enclosure will not interfere with its successful operation. In the case of motor or generator enclosures, leakage that may occur around the shaft may be considered permissible provided it is prevented from entering the oil reservoir and provision is made for automatically draining the motor or generator enclosure (NEMA 4/4X).

WEB FRAME. A built-up frame to provide extra strength, usually consisting of a web plate flanged or otherwise stiffened on its edge, spaced several frame spaces apart, with the smaller, regular frames in between.

WELL. Space in the bottom of a ship to which bilge water drains so that it may be pumped out; space between partial superstructures.

WILDCAT. A special type of cog-like windlass drum whose faces are so formed as to fit the links of the anchor chain. The rotating wildcat causes the chain to be slacked off when lowering the anchor, or hauled in when raising it.

WINCH. A machine, usually steam or electric, used primarily for hoisting and lowering cargo and also for other purposes.

WINDLASS. The machine used to hoist and lower anchors.

WOUND-ROTOR INDUCTION MOTOR. An induction motor in which the secondary circuit consists of polyphase winding or coils whose terminals are either short-circuited or closed through suitable circuits. When provided with collector or slip rings, it is also known as a slip-ring induction motor.

MODULE A - LESSON 1
(57 possible points, 40 passing)

MARINE EXAM

1.   BIBLE (The record for the building of Noah's ark)
2.(a) CENTERLINE
  (b) PLATE
  (c) ABOVE BASE LINE
  (d) AFT PERPENDICULAR
  (e) FRAME
  (f) FORWARD PERPENDICULAR
  (g) CENTERLINE VERTICAL KEEL
  (h) ATHWARTSHIPS
  (i) BOLTED PLATE MAN HOLE
  (j) RAISED WATER TIGHT HATCH
  (k) LENGTH BETWEEN PERPENDICULARS
  (l) LENGTH OVER ALL

3.(a) F
  (b) T
  (c) T
  (d) T
  (e) F
  (f) F
  (g) T
  (h) F
  (i) F
  (j) F
  (k) F

4.(a) G
  (b) D
  (c) A
  (d) F
  (e) C
  (f) E
  (g) B

5.(a) BULKHEAD
  (b) PORTHOLE, PORTLIGHT
  (c) BREADTH
  (d) STACK
  (e) TANK
  (f) ACCOMMODATIONS, STATEROOMS, QUARTERS
  (g) OVERHEAD
  (h) DECK
  (i) LADDER

6.(a) E
  (b) H
  (c) A
  (d) F
  (e) B
  (f) G
  (g) C
  (h) D

7.   CENTER OF VESSEL OR MIDSHIPS

8.(a) ⊏
  (b) T
  (c) ∟
  (d) I

9.   C

10.   NONE

11.(a) RED
  (b) GREEN

MODULE A - LESSON 2
(111 possible points, 78 passing)

ABBREVIATIONS AND INITIALS ON
MARINE DRAWINGS

ABL – Above Base Line
ABS – American Bureau of Shipping
ACCOM. – Accommodations
A & D – Arrangements and Details
AE – Air Escape
AF – Amp Frame
ANT. – Antenna
AP – Aft Perpendicular
ARRGT. – Arrangement
AT – Amp Trip
ATHW. – Athwartships
AUX. - Auxiliary

BALL. – Ballast
BHD. – Bulkhead
BKT. – Bracket
BL – Base Line
B/M – Bill of Materials
BPMH – Bolted Plate Man Hole

CB – Circuit Breaker, Connection
      Box
CC – Control Console
CD – Cofferdam
CL – Centerline
COMP. – Compress, Compressed
COMPR. – Compressor
CONS. – Console
CONT. – Control, Contactor
CONT'D. – Continued
CONTR. – Controller
CSG. – Casing

DETS. – Details
DISCH. – Discharge
DK. – Deck
DO – Diesel Oil
DRNS. - Drains

E & IWD – Elementary and Isometric
      Wiring Diagram

ELEM. – Elementary
ELEV. – Elevation
EMB – Expanded Metal Bulkhead
EMERG. – Emergency
ENG. – Engine
EP – Explosion Proof
ER – Engine Room
EXH. – Exhaust
EXT. - External

FDN. – Foundation
FDRS. – Feeders
FLA – Full Load Amps
FO – Fuel Oil
FP – Forward Perpendicular
FR. – Frame
FW – Fresh Water
FWD. – Forward

GEN. – Generator, General

HP - Horsepower

IB – Inner Bottom
IEEE – Institute of Electrical and
      Electronics Engineers
INBD. – Inboard
INST. – Instantaneous
INT. – Interior, Internal
ISOM. - Isometric

JB – Junction Box

KP – King Post

LBP – Length Between Perpendiculars
LCG – Longitudinal Center of Gravity
L/M – List of Materials
LO – Lube Oil
LOA – Length Over All
LONG'L. – Longitudinal
LT – Light
LTG. – Lighting

Revised 13 April 1999

MODULE A - LESSON 2

ABBREVIATIONS AND INITIALS ON
MARINE DRAWINGS (Continued)

LVP – Low Voltage Protection

LVR – Low Voltage Release

MACH'Y. – Machinery
MAG. – Magnetic
MFLD. – Manifold
MISC. – Miscellaneous
MNS. – Mains
MODS. - Modifications

NAV. – Navigation, Navigating

OT – Oil Tight
OUTBD. – Outboard
OVBD. – Overboard
OVHD. - Overhead

PL – Plate
P/L – Parts List
PLBG. – Plumbing

RED.V. – Reduced Voltage
RPM – Revolutions Per Minute
RWTH – Raised Water Tight Hatch

SHP – Shaft Horsepower
SPT – Sound Powered Telephone
SR – Stateroom
ST – Stuffing Tube
STAN. – Stanchion
SW – Salt Water

TK – Tank
TRANS. – Transverse
TRK. – Trunk
T/S – Toilet & Shower
TT – Tank Top

USCG – United States Coast Guard

VCG – Vertical Center of Gravity
VENT. – Ventilation
VERT. – Vertical
VL – Vertical Ladder

WD – Wiring Diagram
WL – Water Line
WT – Water Tight
WW - Wireway

XFER. – Transfer
XFMR. - Transformer

Revised 13 April 1999

MODULE A - LESSON 3
(82 possible points, 58 passing)

MARINE CROSSWORD

ACROSS
1. CAMBER
3. BILL OF MATERIALS
9. FP (Forward Perpendicular)
10. UPTAKE
13. THWART
15. PASSAGEWAY
16. BUREAU
17. CLEAT
18. PORT
21. BOOBY HATCH
24. CVK (Centerline Vertical Keel)
26. SHIM
30. AP (Aft Perpendicular)
31. PERPENDICULARS
33. EXP (Expansion)
34. AE (Air Escape)
36. FR (Frame)
37. OFFSETS
40. SALT
41. STERN
42. SHELL
44. SHAFT
45. BOLSTER
47. LP (Label Plate)
48. TRANSOM
49. GENERATOR
52. CHAIN
55. BL (Baseline)
56. LB (Life Boat)
57. LIGHT
58. GREEN
59. WT (Watertight)
61. WW (Wireway)
62. STATEROOM
64. TRUNK
68. LOCKER
70. VANE
71. FWD (Forward)
72. OUTBD (Outboard)
73. LADDER

DOWN

2. RPM (Revolutions Per Minute)
3. BINNACLE
4. LOFTSMAN
5. FLUKE
6. ANT (Antenna)
7. ATHWARTSHIPS
8. SPAR
11. ET (Escape Trunk)
12. TRANSVERSE
14. TRUNK
16. BITTS
19. OVBD (Overboard)
20. SCUPPERS
21. BLEEDER
22. HP (Horsepower)
23. CHART
25. KERF
27. STIFFENER
28. LCG (Longitudinal Center of Gravity)
29. PLIMSOLL
32. DOG
34. AFT
35. BREADTH
37. OIL
38. TS (Toilet & Shower)
39. ST (Sounding Tube)
42. STEM
43. LBP (Length Between Perpendiculars)
44. SWASH
46. RED
47. LIN (Linoleum)
49. GALLEY
50. EXH (Exhaust)
51. TIGHT
53. ALTS (Alterations)
54. IB (Inner Bottom)
55. BOW
60. NAV (Navigation)
63. MFLD (Manifold)
65. KP (King Post)
66. INBD (Inboard)
67. XFER (Transfer)
69. EXT (Exterior)

MODULE A - LESSON 4
(18 possible points, 13 passing)

MOTOR CALCULATIONS

Problem No. 1

Circuit Number	Cable	F.L.A.	Circuit Breaker
1-P401	TTNB-500	245	400/300-3
2-P401	TTNB-10	18	100/25-3
3-P401	TTNB-3/0	122	225/150-3
4-P401	TTNB-1	92	225/110-3
5-P401	TTNB-4	61	100/70-3
6-P401	TTNB-8	31	100/40-3
7-P401	TTNB-14	9	100/15-3
8-P401	TTNB-14	6	100/15-3

MODULE A - LESSON 5
(47 possible points, 33 passing)

POWER PANEL FEEDERS

Problem No. 1

Circuit Number	Cable	F.L.A.	Circuit Breaker
1-P403	TTNB-3/0	124	225/150-3
2-P403	TTNB-500	240	400/300-3
3-P403	TTNB-3	65	100/80-3
4-P403	TTNB-5	52	100/60-3
5-P403	TTNB-8	34	100/40-3
6-P403	SPARE	----	400/300-3
7-P403	SPARE	----	225/150-3
8-P403	SPARE	----	100/60-3
P403	(3)TTNB-400	830 (D.L.A.)	1800/1000-3

Problem No. 2

Circuit Number	Cable	F.L.A.	Circuit Breaker
1-P404	TTNB-14	14	100/20-3
2-P404	TTNB-14	14	100/20-3
3-P404	TTNB-14	4.8	100/15-3
4-P404	TTNB-14	2.1	100/15-3
5-P404	TTNB-14	2.1	100/15-3
6-P404	SPARE	-----	100/15-3
P404	TTNB-7	48 (D.L.A.)	100/50-3

Problem No. 3

Circuit Number	Cable	F.L.A.	Circuit Breaker
1-P405	TTNB-14	7.6	100/15-3
2-P405	TTNB-14	7.6	100/15-3
3-P405	TTNB-14	7.6	100/15-3
4-P405	TTNB-14	7.6	100/15-3
5-P405	SPARE	---	100/15-3
P405	TTNB-8	40 (D.L.A.)	100/40-3

MODULE B - LESSON 6
(74 possible points, 52 passing)

LIGHTING CALCULATIONS

Problem No. 1

Circuit Number	Cable	Watts	Amp Load	Circuit Breaker
1-L103	DTNB-14	840	7.00	100/15-2
2-L103	DTNB-14	880	7.33	100/15-2
3-L103	DTNB-14	1020	8.50	100/15-2
4-L103	DTNB-14	980	8.17	100/15-2
5-L103	DTNB-14	350	2.92	100/15-2
6-L103	DTNB-14	300	2.50	100/15-2
7-L103	DTNB-14	1500	12.50	100/20-2
8-L103	DTNB-10	2000	16.67	100/25-2
9-L103	DTNB-14	1500	12.50	100/20-2
10-L103	DTNB-14	1280	10.67	100/15-2
11-L103	SPARE	1066	8.88	100/15-2
12-L103	SPARE	1066	8.88	100/15-2

Problem No. 2

Circuit Number	Cable	Watts	Amp Load	Circuit Breaker
1-L104	DTNB-10	1500	12.50	100/25-2
2-L104	DTNB-10	2000	16.67	100/25-2
3-L104	DTNB-14	600	5.00	100/15-2
4-L104	DTNB-14	800	6.67	100/15-2
5-L104	DTNB-14	1700	14.17	100/20-2
6-L104	SPARE	1320	11.00	100/15-2

Problem No. 3

Circuit Number	Cable	Watts	Amp Load	Circuit Breaker
1-P105	DTNB-14	864	7.2	100/15-2
2-P105	DTNB-14	450	3.75	100/15-2
3-P105	DTNB-14	1870	15.58	100/20-2
4-P105	DTNB-14	696	5.8	100/15-2
5-P105	DTNB-12	2300	19.17	100/25-2
6-P105	DTNB-14	1725	14.38	100/20-2
7-P105	DTNB-14	1680	14.00	100/20-2
8-P105	SPARE	1369	11.41	100/15-2

MODULE B - LESSON 7
(9 possible points, 7 passing)

LIGHTING FEEDERS

Example No. 2

Circuit Number	Cable	Watts	Amp Load	Circuit Breaker
1-P102	DTNB-10	2500	20.83	100/30-2
2-P102	DTNB-14	1500	12.50	100/20-2
3-P102	DTNB-14	528	4.4	100/15-2
4-P102	DTNB-14	696	5.8	100/15-2
5-P102	DTNB-14	696	5.8	100/15-2
6-P102	DTNB-14	864	7.2	100/15-2
7-P102	SPARE	1130	9.42	100/15-2
8-P102	SPARE	1130	9.42	100/15-2
P102	TTNB-6		44 (D.L.A.)	100/45-3

Problem No. 1

Circuit Number	Cable	Watts	Amp Load	Circuit Breaker
L103	TTNB-4*	12782	62 (D.L.A.)	100/70-3
L104	TTNB-8	7920	38 (D.L.A.)	100/40-3
P105	TTNB-5**	10954	53 (D.L.A.)	100/60-3
* TTNB-5 carries 64 Amps, is too close to the DLA, and under the circuit breaker.				
** TTNB 6 carries 54 Amps, is too close to the DLA, and under the circuit breaker.				

MODULE B - LESSON 8
(15 possible points, 11 passing)

TRANSFORMER CIRCUITS

Problem No. 1

PRIMARY CIRCUITS				SECONDARY CIRCUITS			
Circuit Number	Cable	Amps	Circuit Breaker	Circuit Number	Cable	Amps	Circuit Breaker
P415	TTNB-10	30	100/30-3	P215	FTNB-4	70	100/70-4
P416(A)	TTNB-3/0	148	225/150-3	P116 (A)	(2)TTNB-500	541	1600/600-3
P416(B)	TTNB-3/0	148	225/150-3	P116 (B)	(2)TTNB-500	541	1600/600-3
EP407	TTNB-5	59	100/60-3	EP107	TTNB-300	217	None

MODULE B - LESSON 9
(8 possible points, 6 passing)

GENERATOR AND BUS TIE CIRCUITS

Problems

1. Total Generating Capacity is total of all three ships service generators, or 2700 Kilowatts.

2. At sea load cannot exceed total of any two ships service generators, or 1800 Kilowatts. One generator must act as standby.

3. (a) 1800 AF – 1450 LTD – 3 POLE (1443.41 Generator Amps) 115% is 1659.92A
   (b) 1600 AF – 500 LTD – 3 POLE (481.13 Generator Amps) 115% is 553.30
   (c) 1600 AF – 500 LTD – 3 POLE (481.13 Bus Tie Amps) 115% is 553.30

4. (d) Five (5) TTNB-535 (1659.92 Amps @ 115% ÷ 5 = 331.98 Amps per cable)
       (five TTNB-500 @ 329A x 5 =1645A is too low)
   (e) Two (2) TTNB-400 (553.29 Amps @ 115% ÷ 2 = 276.64 Amps per cable)
   (f) Same as for (e).

MODULE B - LESSON 10
(42 possible points, 30 passing)

MOTOR CONTROLLERS

Problem No. 1
1. Steering Gear Motor – Port
2. 30 HP, 440V, 3∅. 60 Hertz, 0.8 Power Factor, 40 F.L.A., 234 Locked Rotor Amps
3. Non-reversing
4. Single Speed
5. 440 Volt heater – 120 Watts (4 Watts per H.P.)
6. Low Voltage Release
7. Full Voltage
8. NEMA 3
9. Steering Gear Room, in sight of motor
10. Splashproof (Controller is located in "wet" space)
11. NEMA 3S
12. "Start-Stop" Pushbutton with Lock on "Stop" (To prevent anyone starting the motor when it is being serviced), Motor "Run" Light (to indicate starting the motor from the remote start in the console)
13. "Start-Stop" Pushbutton, Motor "Run" Light
14. Primary 440 Volts-3 phase, Secondary 115 volts-1 phase (This would not include the motor heater which is 440 Volts –1 phase)

Problem No. 2
1. Ships Service Air compressor No. 1
2. 150 HP, 440 V, 3∅, 60 Hertz, 0.8 Power Factor, 180 F.L.A, 1,080 Locked rotor Amps
3. Non-Reversing
4. Single Speed
5. No heater is used
6. Low Voltage Protection
7. Full Voltage
8. NEMA 5
9. Air Compressor Room, in sight of motor
10. Splashproof (Controller is located in "wet" space)
11. NEMA 3S
12. "Hand-Off-Automatic" Selector Switch (Motor is automatically started and stopped by a pressure switch), Other-Lock on "Off" position (to prevent the motor from being started when under service)
13. "High Pressure Switch", "Low Pressure Switch" (shown in the diagram as a dual switch)
14. Primary 440 Volts-3 phase, Secondary 115 volts-1 phase

MODULE B - LESSON 10

MOTOR CONTROLLERS (Continued)

Problem No. 3

1. Capstan – Main Deck Fwd.
2. 15/7½ HP, 440 V, 3∅, 60 Hertz, 0.8 Power Factor, 21/11 F.L.A, 120/66 Locked rotor Amps
3. Non-Reversing
4. Two Speed
5. 115 Volt – 1 phase heater – 60 Watts (4 Watts per largest HP)
6. Low Voltage Protection
7. Full Voltage
8. NEMA 2
9. Passageway – Second Deck Fwd, not in sight of motor
10. Dripproof
11. NEMA 2/12 (Dripproof, this would not be a General Purpose Enclosure which does not include any drip shield)
12. Motor "Run" Light (to indicate motor has been started from local pushbutton)
13. "Fast-Slow-Stop" Pushbutton (since motor is two speed), with Lock on "Stop" (to prevent motor being started from remotely mounted controller when it is being serviced.
14. Primary 440 Volts-3 phase, Secondary 115 volts-1 phase (which will include the motor heater)

MODULE C - LESSON 14

(25 possible points, 18 passing)

VOLTAGE DROP

Problem No. 1

Circuit Number	Cable Size	F.L.A.	Circuit Breaker	Correction Factor	Voltage Drop %
1-P419	TQNIB-26	40	100/50-3	----	0.93**
2-P419	TQNIB-83	96	225/110-3	0.95	0.17
3-P419	TQNIB-4	14	100/20-3	----	0.49
4-P419	SPARE	----	100/60-3*	----	-----
P419	TQNIB-250	204 (D.L.A.)	225/225-3	1.22	0.09

* Spare Circuit Breaker 4-P419 is rated close to the average of the active trips:
$$50 + 110 + 20 \div 3 = 60$$
** Adding the drop in the feeder (.09) to the largest circuit drop (.93) reveals the largest drop in any cable is 1.02%, which is well under the 6% voltage drop allowed.

Problem No. 2

Circuit Number	Cable Size	F.L.A.	Circuit Breaker	Circuit Length	Voltage Drop %
1-L107	DQNIB-4	3.47	100/15-2	55 (L.C.L.)*	0.96
2-L107	DQNIB-4	5.8	100/15-2	70	2.06**
3-L107	SPARE	4.63	100/15-2	---	-----
L107	TQNIB-4	7.7 (D.L.A.)	100/15-3	30	0.97**

* Load Center Length (LCL) is calculated by adding all lengths and dividing in half.
** Adding the drop in the feeder (.97) to the largest circuit drop (2.06) reveals the largest drop in any cable is 3.03%, which is well under the 6% voltage drop allowed.

Government Institutes Mini-Catalog

PC #	ENVIRONMENTAL TITLES	Pub Date	Price
629	ABCs of Environmental Regulation: Understanding the Fed Regs	1998	$49
627	ABCs of Environmental Science	1998	$39
585	Book of Lists for Regulated Hazardous Substances, 8th Edition	1997	$79
579	Brownfields Redevelopment	1998	$79
4088 ◉	CFR Chemical Lists on CD ROM, 1997 Edition	1997	$125
4089 💾	Chemical Data for Workplace Sampling & Analysis, Single User Disk	1997	$125
512	Clean Water Handbook, 2nd Edition	1996	$89
581	EH&S Auditing Made Easy	1997	$79
587	E H & S CFR Training Requirements, 3rd Edition	1997	$89
4082 ◉	EMMI-Envl Monitoring Methods Index for Windows-Network	1997	$537
4082 ◉	EMMI-Envl Monitoring Methods Index for Windows-Single User	1997	$179
525	Environmental Audits, 7th Edition	1996	$79
548	Environmental Engineering and Science: An Introduction	1997	$79
643	Environmental Guide to the Internet, 4rd Edition	1998	$59
560	Environmental Law Handbook, 14th Edition	1997	$79
353	Environmental Regulatory Glossary, 6th Edition	1993	$79
625	Environmental Statutes, 1998 Edition	1998	$69
4098 ◉	Environmental Statutes Book/CD-ROM, 1998 Edition	1997	$208
4994 💾	Environmental Statutes on Disk for Windows-Network	1997	$405
4994 💾	Environmental Statutes on Disk for Windows-Single User	1997	$139
570	Environmentalism at the Crossroads	1995	$39
536	ESAs Made Easy	1996	$59
515	Industrial Environmental Management: A Practical Approach	1996	$79
510	ISO 14000: Understanding Environmental Standards	1996	$69
551	ISO 14001: An Executive Report	1996	$55
588	International Environmental Auditing	1998	$149
518	Lead Regulation Handbook	1996	$79
478	Principles of EH&S Management	1995	$69
554	Property Rights: Understanding Government Takings	1997	$79
582	Recycling & Waste Mgmt Guide to the Internet	1997	$49
603	Superfund Manual, 6th Edition	1997	$115
566	TSCA Handbook, 3rd Edition	1997	$95
534	Wetland Mitigation: Mitigation Banking and Other Strategies	1997	$75

PC #	SAFETY and HEALTH TITLES	Pub Date	Price
547	Construction Safety Handbook	1996	$79
553	Cumulative Trauma Disorders	1997	$59
559	Forklift Safety	1997	$65
539	Fundamentals of Occupational Safety & Health	1996	$49
612	HAZWOPER Incident Command	1998	$59
535	Making Sense of OSHA Compliance	1997	$59
589	Managing Fatigue in Transportation, *ATA Conference*	1997	$75
558	PPE Made Easy	1998	$79
598	Project Mgmt for E H & S Professionals	1997	$59
552	Safety & Health in Agriculture, Forestry and Fisheries	1997	$125
613	Safety & Health on the Internet, 2nd Edition	1998	$49
597	Safety Is A People Business	1997	$49
463	Safety Made Easy	1995	$49
590	Your Company Safety and Health Manual	1997	$79

Government Institutes

4 Research Place, Suite 200 • Rockville, MD 20850-3226
Tel. (301) 921-2323 • FAX (301) 921-0264
Email: giinfo@govinst.com • Internet: http://www.govinst.com

Please call our customer service department at (301) 921-2323 for a free publications catalog.

CFRs now available online. Call (301) 921-2355 for info.

GOVERNMENT INSTITUTES ORDER FORM

4 Research Place, Suite 200 • Rockville, MD 20850-3226
Tel (301) 921-2323 • Fax (301) 921-0264
Internet: http://www.govinst.com • E-mail: giinfo@govinst.com

3 EASY WAYS TO ORDER

1. Phone: **(301) 921-2323**
Have your credit card ready when you call.

2. Fax: **(301) 921-0264**
Fax this completed order form with your company purchase order or credit card information.

3. Mail: **Government Institutes**
4 Research Place, Suite 200
Rockville, MD 20850-3226   USA
Mail this completed order form with a check, company purchase order, or credit card information.

PAYMENT OPTIONS

❑ **Check** (*payable to Government Institutes in US dollars*)

❑ **Purchase Order** (*This order form must be attached to your company P.O. Note: All International orders must be prepaid.*)

❑ **Credit Card**   ❑ VISA   ❑ MasterCard   ❑ AMERICAN EXPRESS

Exp. ___ / ____

Credit Card No. _____

Signature _____

(Government Institutes' Federal I.D.# is 13-2695912)

CUSTOMER   INFORMATION

Ship To: (Please attach your purchase order)

Name: _____

GI Account # (*7 digits on mailing label*): _____

Company/Institution: _____

Address: _____
(Please supply street address for UPS shipping)
_____

City: _____ State/Province: _____

Zip/Postal Code: _____ Country: _____

Tel: ( ) _____

Fax: ( ) _____

Email Address: _____

Bill To: (if different from ship-to address)

Name: _____

Title/Position: _____

Company/Institution: _____

Address: _____
(Please supply street address for UPS shipping)
_____

City: _____ State/Province: _____

Zip/Postal Code: _____ Country: _____

Tel: ( ) _____

Fax: ( ) _____

Email Address: _____

Qty.	Product Code	Title	Price

❑ **New Edition No Obligation Standing Order Program**
Please enroll me in this program for the products I have ordered. Government Institutes will notify me of new editions by sending me an invoice. I understand that there is no obligation to purchase the product. This invoice is simply my reminder that a new edition has been released.

15 DAY MONEY-BACK GUARANTEE
If you're not completely satisfied with any product, return it undamaged within 15 days for a full and immediate refund on the price of the product.

Subtotal _____

MD Residents add 5% Sales Tax _____

Shipping and Handling (see box below) _____

Total Payment Enclosed _____

Within U.S:	**Outside U.S:**
1-4 products: $6/product	Add $15 for each item (Airmail)
5 or more: $3/product	Add $10 for each item (Surface)

SOURCE CODE: BP01